T0075863

Building Scalable PHP Web Applications Using the Cloud

A Simple Guide to Programming and Administering Cloud-Based Applications

Jonathan Bartlett

Apress®

Building Scalable PHP Web Applications Using the Cloud: A Simple Guide to Programming and Administering Cloud-Based Applications

Jonathan Bartlett
Tulsa, OK, USA

ISBN-13 (pbk): 978-1-4842-5211-6
ISBN-13 (electronic): 978-1-4842-5212-3
https://doi.org/10.1007/978-1-4842-5212-3

Managing Director, Apress Media LLC: Welmoed Spahr
Acquisitions Editor: Steve Anglin
Development Editor: Matthew Moodie
Coordinating Editor: Mark Powers

Cover designed by eStudioCalamar

Cover image designed by Freepik (www.freepik.com)

Distributed to the book trade worldwide by Springer Science+Business Media New York, 233 Spring Street, 6th Floor, New York, NY 10013. Phone 1-800-SPRINGER, fax (201) 348-4505, e-mail orders-ny@springer-sbm.com, or visit www.springeronline.com. Apress Media, LLC is a California LLC and the sole member (owner) is Springer Science + Business Media Finance Inc (SSBM Finance Inc). SSBM Finance Inc is a Delaware corporation. For information on translations, please e-mail editorial@apress.com; for reprint, paperback, or audio rights, please email bookpermissions@springernature.com.

Apress titles may be purchased in bulk for academic, corporate, or promotional use. eBook versions and licenses are also available for most titles. For more information, reference our Print and eBook Bulk Sales web page at http://www.apress.com/bulk-sales.

Any source code or other supplementary material referenced by the author in this book is available to readers on GitHub via the book's product page, located at www.apress.com/9781484252116. For more detailed information, please visit http://www.apress.com/source-code.

Printed on acid-free paper

Cloud computing is really a no-brainer for any startup because it allows you to test your business plan very quickly for little money. Every startup, or even a division within a company that has an idea for something new, should be figuring out how to use cloud computing in its plan.

—Brad Jefferson, CEO of Animoto

Table of Contents

About the Author

Jonathan Bartlett is a technical lead at ITX, where he leads a team of programmers doing work as diverse as building online registration systems, modeling the physics of oil flow through a well, and developing augmented reality body evaluation applications. Jonathan's work focuses on cloud development, Ruby on Rails, data modeling, and iOS development. Jonathan is patently awful at building user interfaces, and is thankful for the large number of more creatively minded coworkers who cover for him on a daily basis.

Jonathan has been educating programmers for well over a decade. His first book, *Programming from the Ground Up,* is an Internet classic and was endorsed by Joel Spolsky, co-founder of Stack Exchange. It was one of the first open source books and has been used by a generation of programmers to learn how computers work from the inside out, using assembly language as a starting point. Recently, Jonathan released *New Programmers Start Here* which is focused on teaching brand new programmers about computers, the Internet, and JavaScript programming, based on his experience teaching programming to high-school and college students. Additionally, Jonathan has written several books on the interplay of philosophy, math, and science, including *Calculus from the Ground Up, Engineering and the Ultimate,* and *Naturalism and Its Alternatives in Scientific Methodologies.*

Jonathan also writes developer-focused articles for a number of technology web sites. His articles can be found on IBM's DeveloperWorks web site, Linux.com, and Medium.com. Jonathan is currently writing technology articles for a more general audience at MindMatters.ai.

Jonathan also participates in a variety of academic work. He is an associate fellow of the *Walter Bradley Center for Natural and Artificial Intelligence*. There, he does research into fundamental mathematics and the mathematics of artificial intelligence. He also serves on the editorial board for the journal *BIO-Complexity,* focusing on reviewing information-theoretic papers for the journal. Jonathan served as editor for the book *Controllability of Dynamic Systems: The Green's Function Approach,* which received the RA Presidential Award of the Republic of Armenia in the area of "Technical Sciences and Information Technologies." He also spends time teaching at a homeschool co-op through Classical Conversations.

Jonathan is married to Christa. They have been together for over 20 years and have had five children, though they lost two of them due to a genetic illness.

Acknowledgments

This book was written based on a number of experiences moving clients from traditional hosting environments to cloud-based hosting environments. The material here started as some notes that I put together for other people on my team and eventually blossomed as the book that you see here.

I would like to thank my clients for giving me the opportunity to work with them on exciting projects and for allowing me to explore and learn new systems on their behalf. I would also like to thank my employer, New Medio (now part of ITX), for always having great clients to work with and always putting up with all of my little side projects. A number of companies look down upon employees doing side projects. They have always been encouraging to me in my endeavors, whether it is teaching science to homeschool co-ops, going to seminary to study theology, organizing seminars on obscure subjects, or my steady stream of books. Thanks especially to Adam Nemec for making all of these things possible. I doubt I could do them from anywhere else.

Next, I want to thank my early test readers, Tavo Soto, Charles McNamara, Garret Wilson, and Cara Waken, for taking the time to read through the book and test out the different examples to make sure they functioned properly. I also want to thank David Roesch and Alex Fornuto for their help with many technical details in early drafts of this book. Finally, I want to thank the editorial staff at Apress for their great work helping me to make this book perfect. It is a much stronger book because of your help.

CHAPTER 1

Introduction

"The cloud" is the new technology buzzword. It seems like everyone is talking about "moving to the cloud," but few people really know what that means, least of all how to take advantage of it. With every technology shift, there are people who think that the new technology will solve all of their problems, without even taking inventory of what those problems are. Many technologies make promises, and some of those are even true, but for them to come true, you have to use the technology in the right way.

This book is primarily for developers who want to start moving their apps to the cloud, and want to know how to get started and different options available to them. Secondarily, this book is for managers who want to be conversant in cloud technology and ideas, in order to better understand the options available and how different development decisions will affect them. This book has a focus on PHP, but, even if PHP is not your intended deployment language, this book will tell you what you need to know to architect a cloud application in any language or platform.

This book will talk about several different cloud options, but will focus on developing for Linode and similar clouds, the reasons for which we will see in Chapter 2.

Please note that this book makes many comparisons between specific infrastructure vendors. I do not represent any particular company, nor do I warrant that these comparisons are foolproof or permanent. They are, however, the results of my experience and my knowledge at the time of writing, and I have attempted to do all due diligence to provide factual

J. Bartlett, *Building Scalable PHP Web Applications Using the Cloud*,
https://doi.org/10.1007/978-1-4842-5212-3_1

information as best I can. Nonetheless, while the general considerations discussed in this book are unlikely to change, the specific vendors, products, and how they measure up *are* likely to change over time.

Additionally, while the general outlines of the tasks for setup, programming, and configuration are likely to remain the same for a long period of time, the specific steps and screenshots may differ from this book as products and platforms change. Nevertheless, the steps presented here should remain a reliable guide for what types of things you should expect from different cloud vendors.

1.1 Prerequisites

This book has very few prerequisites. Even if you aren't familiar with the specific tools we are using, this book has enough step-by-step instructions that you should be able to follow it fairly easily.

Since it is a book on web applications, it assumes that you are familiar with the absolute basics of HTML, CSS, and the basics of how the Internet works (i.e., domain names, IP addresses, etc.). If you are not familiar with these things, you should put this book down and pick up a copy of my earlier book, *New Programmers Start Here*. *New Programmers Start Here* doesn't hit every topic you need to know, but if you understand its concepts, you should be able to work through most of the examples in this book.

The code in this book is based on PHP, but the code is simple enough that you should be able to follow it no matter what language you are familiar with. PHP was chosen because it is easy to write short, understandable web applications with PHP. For instance, in my job, I almost always program in Ruby on Rails. However, it is almost impossible to understand a Rails application without already being familiar with the entire Ruby on Rails system. PHP, on the other hand, is less "magic" than Rails, but that makes it more obvious to the reader what is happening.

With PHP, especially at the level which this book uses it, I'm confident that anyone who has experience with any programming language will be able to follow what is happening.

This book also presumes that you have a basic familiarity with databases and SQL. This is not an absolute requirement, but the database code itself is left largely unexplained. However, the SQL code should be self-evident to anyone with even a passing knowledge of SQL.

Finally, this book uses Linux as the operating system of choice. However, even if you don't know anything about Linux, this book gives you every command you need to type, so you don't actually have to know anything about Linux to use this book. Appendix A has a list of common Linux commands for reference if you want to know more.

The reason that Linux was chosen was because it is a very easy operating system to install and use in a cloud environment. While it is possible to use Windows in a cloud environment, Linux was built from the ground up for this type of application. The command line, while seemingly arcane, actually makes server management fast and painless.

Additionally, because Linux, PHP, and PostgreSQL (our database system) are all free software, there are no licensing considerations for their use. You just install what you need and move on with your life. You don't need to worry about who you need to pay when. You don't need to worry about whether your usage matches your license for the product. You don't need to worry about someone auditing your business to verify compliance. With free software, software is just a tool—you can install it and forget it.

I don't have a problem with proprietary software in principle, but, because it introduces a ton of extra management issues, I try to avoid it if possible. I am not against people being paid for their work, but I do take a principled stand against never-ending, needless headaches, which is what you wind up with when you rely too much on badly licensed proprietary software.

This book focuses on the CentOS 7 Linux distribution because CentOS tends to be a fairly stable and robust Linux distribution that is supported by most vendors, and many other Linux distributions use CentOS as a starting point.

1.2 Typographical Conventions

This book uses a few, simple typographical conventions you should be aware of. When referring to code, typed commands, URLs, or any other text meant for typing, the book uses a font that looks like this: `type me here`. Names of programs which are the same as their command name (which you would type in to use) are also written in that way, as are usernames and filenames.

However, many things that you type will need to be replaced using your own values, such as the IP addresses for the machines that you deploy. These things that need to be replaced are given by words with all uppercase letters. For instance, to go to your own web site, you would type into the browser `http://YOUR.DOMAIN.HERE/`, where `YOUR.DOMAIN.HERE` refers to your own web site's domain name. The meaning of these and what they should be replaced with are described in the text.

Note that PHP actually has a few default variable names that are all uppercase that we will be using. Therefore, `$_GET`, `$_POST`, and `$_FILES` are all real variable names in PHP and should not be replaced.

This book describes how to use a variety of third-party services. When describing the menus, buttons, and field names of user interface elements on various web sites, these items will usually be placed in double quotes. For instance, to print out a file from most programs, you would go to "File" and then click "Print."

1.3 Typing or Downloading the Code

This book focuses around a simple guestbook application, building and rebuilding it using different application architectures. The entire code for the application is in this book. Honestly, I think it would do you a lot of good to type out all of the code in this book yourself. There isn't very much, and it would help you think about what you are building. However, I understand that this can get tedious when you are just trying to work the examples, and having complete, working code is very helpful when trying to solve problems.

Therefore, all of the code for every variation of the application we cover in this book is available for download from `www.github.com`. Each branch in the repository represents a different modification to the application in the book. Using the repository, the only changes you would need to make are the environment-specific changes mentioned in the text of the book, such as setting the IP address of the database server. These are also in the `README` file for each branch.

You can download the code at

```
https://github.com/johnnyb/cloud-example-application
```

If you do type in the code yourself, pay close attention to where the spaces are used in the program. The code is written exactly as it should be typed, with line breaks where they should be unless indicated elsewhere by the text. Putting spaces and line breaks in the wrong place or leaving them out where they should be is a surefire way of messing up the code.

CHAPTER 2

What Is the Cloud

There is a little bit of confusion over what it is to "be in the cloud." Some people even mistakenly think that just being on the Internet is utilizing cloud technology.

The essential difference between cloud technology and other types of Internet hosting is that a cloud service offers at least the ability to scale your application quickly. Historically, developers would deploy their web application to fixed servers that they purchased or leased at a specific facility. It was possible to get more equipment, but this always involved quite a bit of time and effort. Oftentimes, the developer would have to put together a large capital outlay for the servers, purchase them, configure them, and deploy them. This process could take weeks or even months. Even when purchasing direct from a hosting company, the process of putting together a quote and getting everything up and running could take more than a week.

The promise of the cloud is that, rather than having to go through the process of physical setup for new servers, additional capacity can be granted either instantaneously or at least within minutes or hours rather than days, weeks, or months. With some solutions, you don't even have to worry about machines—the cloud solution automatically scales your application across as many machines as you need. With others, it is just a few clicks to request, image, and boot up a new machine that is a replica of some existing machine and then add it to your web application.

In any case, the core idea of the cloud is instant, automated scalability and flexibility.

J. Bartlett, *Building Scalable PHP Web Applications Using the Cloud*,
https://doi.org/10.1007/978-1-4842-5212-3_2

Many cloud providers even provide access to more than one data center. Do you want a set of servers in America and a set of servers in Europe? No problem. With just a few clicks, you can get it done.

2.1 Infrastructure as a Service

Cloud computing is often confusing because there are several different *types* of cloud computing, each of which has its own benefits and drawbacks. The difference between the types of cloud services is mostly based on the level of abstraction that is being offered.

The most basic type of cloud service, which this book will focus on, is called *Infrastructure as a Service,* abbreviated IaaS. IaaS means that you can purchase and deploy pieces of infrastructure (servers, load balancers, firewalls, etc.) with just a click of the button. This is done through server virtualization technology.

Server virtualization allows an IaaS service company to deploy a single, very large server and split it up into multiple, smaller servers. Each server runs a *hypervisor* program, which allows the company to split up the server quickly and automatically. A company might take a computer with 16 processors and 64 gigabytes of RAM and split it up into 4 virtual machines, each with 4 processors and 16 gigabytes of RAM.

Doing this has three advantages. The first is space. By buying the biggest machine, the IaaS company gives themselves the most computing power per rack unit, which is a valuable commodity in server rooms. Therefore, by buying a single, big machine and splitting it into four smaller machines, they have only used a fourth of the space that they might otherwise have done.

The second is cost per CPU core. By packing so many CPU cores and so much memory into a single server, their cost per CPU core and cost per gigabyte of memory go down. Therefore, by buying larger computers, they can deliver a lower cost per CPU core to the user who buys the smaller, virtual servers.

The third advantage, however, is the most important—manageability. In order to support virtualization, a small part of each machine is dedicated to the hypervisor—the small operating system that manages the other virtual machines on the server. Because the machines are virtual machines, and not real hardware devices, they are incredibly easy to manage. An operator can send a command to a hypervisor, and it can instantly set up a new virtual machine, clone a new boot disk, and start up the new virtual server in minutes.

Historically, if I wanted a new server, I would have to buy the server, then, while being physically at the console, would have to install the relevant operating system and system software, and finally carry the server to the rack where I plug it in and turn it on. I have to make sure the network is plugged in and configured. I might have to configure the BIOS to allow for keyboard-disconnected operation. If there is a problem with the machine, I have to make a backup of the machine, find a new physical machine, copy the backup to the new machine, install the new machine, and physically swap it out with the old machine.

With virtualized machines, I can let the hypervisor take care of all of that. All I need are enough extra servers running hypervisors so that when I need a new machine, I can just tell a hypervisor to create a new virtual machine for me, and where the copy of the disk is that I want to use.

What's even better is that with most IaaS platforms, you don't even need to worry about hypervisors and capacity. The IaaS vendor does all of that for you. The IaaS vendor provides a point-and-click interface that allows you to boot up virtual servers just by logging in to your web management console. You tell it how big of a machine you want (i.e., number of CPU cores and size of memory), and it will allocate a server for you. You tell it what you want on it (either a base operating system or a clone of an existing machine), and it will copy that to its boot disk and boot it up for you. Voilà! You have a new server ready to go.

Additionally, with most IaaS services, if there are physical hardware issues, the service will take care of it for you. If a severe hardware problem is detected, they will simply turn off your server, migrate it to a new one, start it back up, and send you an e-mail letting you know what happened. If it is a less severe problem, some vendors will give you a notification, requesting that you push a button to perform the migration at a time that is most convenient for you.

Even better is the pricing model. Most IaaS vendors have an option for *hourly* pricing. Do you need a machine, but just for a few hours? With Linode, you can get a 32-core and 64 gigabyte machine for well under $1.00 per hour!

In Chapter 3, we will look at the specific steps required to set up a cloud server.

2.2 Platform as a Service

Another option for scaling applications is known as *Platform as a Service,* which is abbreviated as PaaS. With PaaS, rather than giving you bare machines and letting you run whatever you wish, a PaaS *defines* a platform for you to run your application on. The PaaS vendor manages the entire platform, and you just have to worry about your application.

For instance, Heroku is a popular PaaS vendor. Heroku handles all of the server administration and maintenance. You don't even get to log in to their machines! You simply push the code to them, and they deploy it on their machines for you. Other common PaaS vendors include Google App Engine, Windows Azure, Amazon's Elastic Beanstalk, and OpenShift.

With PaaS, you choose how many "workers" to use, and the system will distribute that work across however many machines are needed (a worker is just an active process, and each PaaS vendor has their own term for it). Thus, the PaaS vendor takes care of the *platform* (hardware, operating system, installed applications), and you just manage the application code.

This sounds great in theory—you no longer have to manage servers at all! It is all done for you automagically behind the scenes. However, the reality is that PaaS platforms are not as transparent as they appear, and PaaS vendors tend to price their value addition fairly high.

2.3 Docker

A new technology in the field generating a lot of buzz is Docker. Docker apps are kind of a midway between IaaS and PaaS. A Docker app is an image containing, essentially, a full installation of all of the programs needed to run your application.

They can be managed, deployed, and scaled similar to PaaS, where you can just tell the service how many you want running, and it will take care of deploying your image to the right places, but you have a much larger degree of control, similar to IaaS.

One interesting part of Docker is the ability to "link" different services together. Essentially, you put names on the different types of services that a Docker container provides or needs, and then can use those names to tell your services how to find each other. However, this is mostly only an issue during the initial setup for your application, after which it matters little whether you linked your application components together manually or with a fancy tool.

Docker itself is a technology, not a vendor. There are many vendors which allow you to deploy Docker apps, including Docker Inc., the company responsible for the Docker technology.

2.4 Why Choose IaaS

This book concentrates on IaaS cloud models for a number of reasons. The first is that IaaS is very flexible. No matter what type of workload you want to run, no matter what platform you want to run, IaaS just gives you bare servers. What you do with them is up to you. This also means that you aren't locked in to a specific vendor's systems.

The second is that IaaS works fairly predictably. While there are differences, the way that most IaaS vendors operate is fairly similar. You pick a box size, you say what you want on it, and you push the button. There is a lot of variability of the quality, cost, and flexibility of those services, but they are all fairly similar.

With PaaS, your choices are much more limited. First of all, you are limited to what platforms your vendor offers. If you program in Ruby on Rails, you are limited to only Ruby on Rails PaaS. While most PaaS vendors have opened up their systems to quite a number of different application servers, there are still limitations. Second, you must write your code to match the way they have their platform configured. This usually involved little or no local file storage, certain specific types of databases to connect to, only certain allowed platform options or extensions, and little to no flexibility on how the server is configured. Sometimes that's fine, but sometimes it is too constraining.

Additionally, PaaS is usually too opaque. Access to server logs is often difficult, and debugging troubles that are server-specific is nearly impossible. Sometimes PaaS vendors have tools to help, but they are nothing like being able to debug directly on the machine having the problem. This opaqueness can sometimes lead to devastating results. For instance, if you purchase a PaaS database system, and somehow your database gets corrupted, your options are extremely limited. You basically have to call the company and beg for help. Some companies are very responsive in this regard, but it makes me really nervous.

Also keep in mind that in order for a PaaS system to work, they have to keep upgrading their systems. In fact, that is what you are paying them to do. However, there is no guarantee that an upgrade they make tomorrow won't also accidentally trash your application. Perhaps the upgrade was necessary, but that isn't a decision that you get to make.

Similarly, using a PaaS vendor means that you have to upgrade on *their* schedule. If they decide to deprecate a piece of their infrastructure, you have to rewrite your code to work around this. If they decide that the version of a piece of software you use is out of date, you have to rewrite your code to use the new version. In short, PaaS means that you lose control of your technology, while IaaS means that you have full control. While certain hassles are removed by using PaaS, they are usually compensated for by the hassles they introduce.

Finally, PaaS is usually very expensive. For example, for four CPU cores on Heroku (a common Ruby on Rails PaaS vendor), it costs $100 per month. For a 4-core machine on Linode, the cost is only $40 per month, and the Linode machines are faster. I have found that most PaaS vendors charge around three times as much as a good IaaS vendor for equivalent performance. The PaaS vendors do more for you, but that is only worthwhile if you don't possess any systems management experience in your developer lineup, and you will often wind up paying that cost anyway in keeping up with the PaaS platform.

If your organization has the know-how to set up and maintain servers (and this book provides a good start to learning how to do that), IaaS is currently the easiest, most flexible, fastest, and least expensive way to take advantage of cloud technologies, and it doesn't lock you in to a specific vendor. If you decide that you really want to use a PaaS-like system, just know that there are many open source PaaS systems that you can run on top of your IaaS servers.

2.5 Choosing an IaaS Vendor

There are many considerations that go into choosing an IaaS service or vendor. The most important consideration is the reliability of the service. It doesn't help to have a magnificent web application if the service is down. It doesn't make your company money in the long term if you can't get to your data. Therefore, the reliability of the service should have significant impact on your decisions.

This also extends into their ability to resolve tickets. Every service will have problems at some time or another. If a company cannot respond to or resolve tickets in a timely manner, then you shouldn't have production systems sitting on them.

The next main factor is the price/performance ratio. Many of the early cloud infrastructure companies placed an extremely high value on the flexibility of cloud computing, with the result that nearly every cloud solution was at a horrendously high cost. Amazon Web Services (AWS) has a cloud computing service known as EC2 that is a great example of this. As mentioned earlier, a Linode 4-core machine with 8GB RAM is $40/month. A similarly specified machine on EC2 (the `c5.xlarge` machine) is $122/month. Historically, EC2 has tended to be much slower than Linode even on the same specs. In the early days of cloud computing, EC2 was one of the only big players in the field, and to get the flexibility, they made you pay a high price. Personally, at the price EC2 is asking for, I would prefer to go outside of the cloud and do a traditional hosting setup where I lease physical servers. For me, the flexibility isn't worth the price that AWS requires.

On the other side of the scale, there are cloud services whose prices are impossibly good. That is, you can recognize from the price that there is no way that they can deliver reliable long-term service at that price. Either the company will go out of business, the company will have to raise prices, or the service will eventually be oversubscribed and degrade to the point of unusability. A great example is CloudAtCost (`www.cloudatcost.com`). With CloudAtCost, you pay a one-time fee and get to keep the server *forever*. For instance, for $70, you get 2 CPU cores and 1GB of RAM. But that is *not* a monthly cost. That is a *one-time-only* cost! As I said, it is an unbelievably good price. They do charge a $9 yearly account maintenance price, but that is the same cost no matter how many servers you have (it was probably instituted so they could shut off servers that were no longer maintained).

There are some things that such a service might be good for. If you wanted a development server, for example, it doesn't matter if the network goes down every once in a while, or that the technicians aren't quick to answer. And, if one day they shut their doors, you aren't out a whole lot. But I certainly wouldn't bet my business on such a service. If you want to see other services that are at impossibly cheap prices, see `www.lowendbox.com`.

The final factor is flexibility. It is one thing to be able to point and click to set up a new machine, but what if you couldn't store disk images and had to rebuild a machine every time you launched one? This would be a painstaking process, and you would lose one of the main benefits of cloud computing. This area is one in which AWS shines. AWS consists of not only its cloud computing services (EC2), AWS also provides configuration, control, and automation into nearly every aspect of cloud infrastructures. AWS provides scalable storage solutions, video transcoding services, scalable databases, message queueing services, and search services. This gives your cluster an external monitoring service which automatically spawns new servers to handle the load as it increases, and removes servers from your cluster and powers them down as the load on your network decreases. In other words, the extra services that come with AWS are nearly limitless.

At the end of the day, however, the flexibility of AWS is outweighed by the terrible price/performance ratio of EC2. Thankfully, however, AWS's better services (such as S3 and CloudFront) can be used individually, even if you use another provider as your primary IaaS vendor. We will discuss how to do that in later chapters.

If you haven't already guessed, my preference for a cloud vendor is Linode (`www.linode.com`). Their price/performance ratio is unbeatable. The reason for this is threefold. First, the equipment is newer. Second, they only use SSD drives (i.e., solid state—no spinning disks), which are generally an order of magnitude faster than regular disks. This is probably one of the most important aspects to Linode's performance. Third, they have implemented controls to prevent "noisy neighbors."

A noisy neighbor is a virtual machine that is on the same physical server as your machine, but is using all of the I/O resources of the computer. On IaaS platforms, you don't get to choose (or even know) who else is sharing the same physical hardware. Therefore, using a service that prevents resource hogging is very important. Linode has implemented a number of controls to prevent any individual virtual machine from overusing resources. This is good not only for the performance of your own virtual server but also because it means that you don't have to worry yourself about being a bad neighbor to someone else!

While Linode does not have the flexibility of AWS, the flexibility is hardly missed. Most of the important things that you want to do with AWS are extremely simple with Linode, and the added features of AWS make AWS's learning curve steep and confusing. By the time someone spends the extra money to get the flexibility in AWS (for instance, creating a system to automatically boot new machines in response to load), they could have spent the same money in Linode giving their cluster enough capacity so that it won't matter.

There are other services similar to Linode—DigitalOcean is often considered a comparable service at a comparable price. However, my experiences with Linode have been sufficiently positive that I haven't felt the need to research them all. Linode provides what I need, has an easy-to-use interface, and does it at a great price. Therefore, this book will focus on developing cloud applications on Linode.

When choosing a vendor, it is also important to read their terms of service carefully to make sure that your intended use is compatible. Not only are some workloads not allowed on certain services (i.e., some services prohibit mass e-mail marketing), many services have limits on how fast you are allowed to scale your network. This is mostly for abuse prevention, but it is important to find out about ahead of time. In any case, you may want to reach out to potential cloud vendors and make sure that your planned usage matches their service guidelines before settling on one.

2.6 Some Important Terminology

Before we go further, I want to clarify some of the terminology that is used in the book. With IaaS, what you are renting are basically machines. They are *virtual* machines (i.e., partitions of real machines), but nonetheless they can themselves be considered machines in the abstract. In cloud computing lingo, they are often referred to as *nodes*. Since they provide services to other machines and/or users on the network, they are also considered to be *servers*. Because of this, another term often used for nodes is VPS—virtual private server. In this book, the terms machine, node, and server are used fairly interchangeably, though the specific term chosen is usually based on how you are using it at the moment.

A *cluster* is a group of nodes that work together. Thus, we are building a cluster in the cloud. A service that is built to be scalable in the cloud is called a *cloud cluster*, or just a *cloud*.

The term *scalability* is an important term in cloud computing. Scalability is different than just performance. Performance refers to either efficiency or raw speed. Scalability refers to the ability of the system to expand its processing power quickly. For instance, a program that can handle 2,000 requests per second might be considered high performing, but if that program's speed cannot be increased by adding another node, it is not scalable.

On the other hand, even a low-performing program might be scalable. If my program only handles 1 request per second, but every node I add increases the number of requests it can handle, then my program is scalable. If I scale this low-performing program up to 4,000 nodes, it will be able to handle more requests than the previous case which handled at most 2,000 requests per second. The whole cluster (with its 4,000 nodes) might now be considered high performance based on the throughput of the whole cluster even though its individual parts are not high performance.

Usually, scalability depends on your ability to *parallelize* tasks. Parallelization refers to the ability of tasks to run independently of each other, even on other machines. The more that two tasks need to coordinate with each other, the less parallelizable they are. In designing applications for scalability, you need to take particular note of the way in which different parts of the system require coordination, and aim to minimize or remove the impact of these coordination points.

Hopefully, the things that you build are *both* high performance (even for a small number of nodes) *and* scalable (so that adding more nodes increases performance). Making an application perform well is about looking for slow code and replacing it with fast code or rewriting your application to not need the slow code. Making an application scalable is about looking for *bottlenecks* which prevent parallelization and replacing those bottlenecks with code that can scale to multiple nodes or reworking your architecture so that the bottleneck doesn't arise in the first place.

The focus of this book is on scalability, though performance is also considered.

SKIPPING THE SERVER ROOM

Here is an amusing story of things that I don't miss about dealing with server rooms. In the late 1990s, I once worked as a programmer/system administrator for a company that hosted its own servers. These servers were not in regular server racks, but rather on essentially metal wire shelving. The primary external-facing server was on the top shelf.

We needed some extra drive space on this server, so I had installed an external drive pack on the server. At the time, the main mode of connecting external storage to servers was called SCSI (pronounced "scuzzy"), and it was fraught with problems. After some time, there started being a lot of problems with the server accessing the drive pack, so I went to troubleshoot.

Since the server was on the top shelf, I stood on top of a stool to work with the machine. I spent about a half an hour trying to figure out what was wrong. Realize that this was happening with our main external server, so every minute counted, because our (very active) web site was having problems.

So, I managed to move everything off of the new drive packs, then disconnected the drives to take them back to my office to see what the problem was. However, I had spent so much time standing on the stool messing with the computer, I forgot that I was, you know, standing on a stool. So, I grabbed the drive pack and just walked out of the room, or so I intended. Thankfully, neither I nor the drive pack was injured by the fall, but my pride certainly was.

As fun as stories such as these are to reminisce over, I'm glad that those days are behind us.

CHAPTER 3

Setting Up a Cloud Server

As with most cloud providers, setting up a cloud server on Linode is extremely simple. This chapter shows you the basic process to get something up and running, with screenshots.

3.1 Creating Your Virtual Server

Hopefully, if you are a web developer or manager, you can sign up for the Linode service without too much trouble. It requires a credit card up front, but you will find the costs of running everything in this book are likely less than the book itself, assuming that you turn off your services when you are done with them.

Before moving ahead, go ahead and create your account with Linode now.

After you sign up and sign in, Linode brings you to your dashboard, which should look something like Figure 3-1.

J. Bartlett, *Building Scalable PHP Web Applications Using the Cloud*,
https://doi.org/10.1007/978-1-4842-5212-3_3

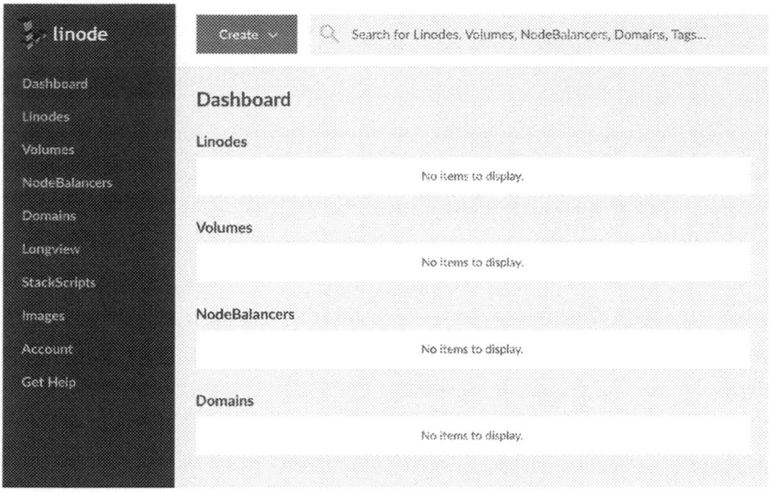

Figure 3-1. *The Linux Dashboard*

You currently have nothing set up, so your dashboard is pretty empty. To get started, click the "Create" button. Linode calls their virtual servers "nodes" or "Linodes," so choose "Linode" to create a new machine.

Linode will then ask you several questions to help you configure your node for use. While there are any number of good choices, use the ones here in order to be able to follow along with the book:

1. Under "Choose a Distribution," pick "CentOS 7."
2. Under "Region," it doesn't matter what you choose, but you have to pick the same one each time for your servers to talk to each other. This book will use the "Dallas, TX" facility.
3. Under "Linode Plan," the cheapest one, which is sufficient for our purposes, is the "Nanode 1GB" plan. At the time of this writing, this plan costs less than 1 cent per hour to operate.

4. Under "Linode Label," we will call this machine `template_node`.

5. You can ignore the "Add Tags" section. Tags are useful for grouping machines together when you have a lot of them.

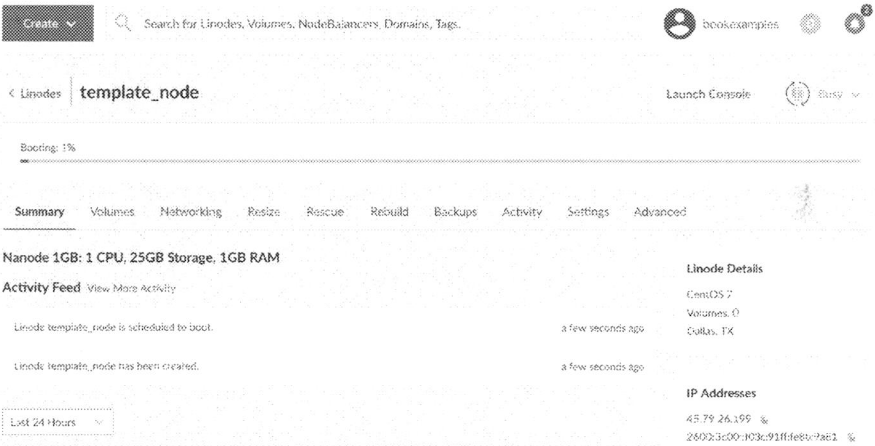

Figure 3-2. *Your New Node's Dashboard*

6. Under "Root Password," add a password for this machine. Be sure the password is secure, as there are a lot of hackers that just go around attempting various passwords for root accounts. It is not unusual to have 50,000 such hack attempts each month.

7. For the moment, you can leave "Optional Add-ons" untouched. We will deal with backups and private IP addresses later on in the book.

After setting all of these, hit the "Create" button, and Linode will start building your machine. Linode will bring you to a dashboard for your machine which will have, among other things, a progress bar and an "Activity Feed" (see Figure 3-2). When the progress bar is done, you are now the proud owner of a new server in the cloud!

CHOOSING A SERVER

Linode categorizes their servers by (a) the kind of workload they will be serving and (b) how much RAM the server comes with. The workload types are "Nanode" (super-small and super-cheap instances), "Standard" (balanced CPU/RAM), "Dedicated CPU," and "High Memory" (there is also "GPU," but those are for a completely different type of cloud computing than is considered here).

Under each workload, the servers are named by the amount of RAM they ship with. A "Linode 4GB" comes with, you guessed it, 4 gigabytes of RAM. It will also specify how many virtual CPUs and how much disk space it comes with as well. Generally, the amount of memory goes up faster than the number of cores, which makes sense considering the fact that memory is often a much stronger limiting factor than CPU power. They also each have differing amounts of disk space, but I consider the disk space a minor consideration, since Chapter 8 will show how to set up a service with infinite available disk space.

When choosing a production server, for reasons that will become more clear going forward, I usually choose a fairly large size for databases (since they are harder to replicate) and low to midrange sizes for web servers (since I can easily add capacity by adding more servers).

3.2 Logging In and Looking Around

So, you have a machine, but where is it and how do you access it?

Click the "Networking" tab on your dashboard. It should look something like Figure 3-3. Under "Access" there is an area titled "SSH Access." This has the command you need to type into your command line in order to log in. Use the password you created for the node to log in.

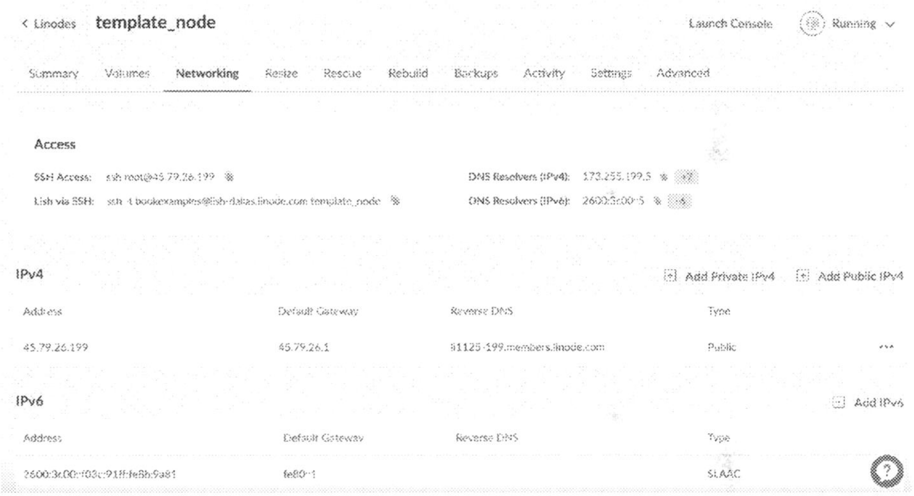

Figure 3-3. *The Networking Tab*

> **☞ COMMAND LINE? WHAT IS THAT?**
>
> The command line is the old-school way of accessing computers. Long before there were pretty graphical interfaces, people interacted with their computers by typing. For many things, *especially* for system administration–related tasks, the command line remains the best way to manage systems.

If you don't have any experience with the command line, don't worry! This book doesn't assume that you have expertise in this area, and will walk you through each step. If you are just trying to figure out how to *get to* the command line, here is what you do in each major operating system:

- **Windows 10**: Windows actually has *two* command-line systems. The older "Command Prompt" (`cmd.exe`) and the newer PowerShell. Just click the Windows icon and type `PowerShell` to get started. You will need at least the April 2018 update in order to run the commands here without further setup.
- **MacOS X**: Every Mac comes with an application called "Terminal." You can find it with Spotlight search, or you can go to "Applications," then "Utilities," and find it there. I would suggest that you add it to your Dock, as you will probably use it a lot.
- **Linux**: Every Linux distribution has a command-line program installed, usually named "Terminal" or "Bash Prompt" or something along those lines.

When you start the command line for the first time, it will show you some text, followed by a blinking cursor. Now you are ready to start typing commands!

In case you are new to Linux, `ssh` is an incredibly handy tool that allows for a remote, secure connection to your server's command line. That is, you can `ssh` into your machine, and it works just as if you were logged in to the console. Additionally, the connection is encrypted, so you don't have to worry about anyone eavesdropping on you or stealing your password. `ssh` is installed by default on every major operating system, so you should already have it installed. If you are using an old version of Windows, you may have to download a separate ssh application, such as PuTTY, which can be freely downloaded from `www.putty.org`.

To log in to your machine, simply open up a command line and type in the command listed under the "SSH Access" section. It should say something like `ssh root@MY.IP.ADRESS.HERE`, where `MY.IP.ADDRESS.HERE` is the IP address of your Linode. Because `ssh` has not seen this computer before, it will likely warn you that the authenticity of the host can't be established, and ask whether or not you want to continue connecting. Just answer `yes`. It will only ask you this the first time, as `ssh` will remember the remote computer. Then, for the password, put in the password that you set when setting up your machine. Figure 3-4 shows what this will likely look like.

```
Jonathans-MacBook-Pro-4:~jonathan$ ssh root@45.79.26.199
The authenticity of host '45.79.26.199 (45.79.26.199)' can't be es-
tablished.
ECDSA key fingerprint is SHA256:1IqKL71RoknJwjKYLGc.
Are you sure you want to continue connecting (yes/no)? yes
Warning: Permanently added '45.79.26.199' (ECDSA) to the list
of known hosts.
root@45.79.26.199's password:
[root@li1125-199 ~]#
```

Figure 3-4. *Logging In from the Command Line*

You are now logged into your machine!

That final line is known as the "command prompt." It gives basic information about the current state of your session. `root` is your username. This is the name of the administrative user on Linux. `li1125-199` (or whatever is after the `@` symbol) is the name of your machine. Finally, ~ tells you what directory you are in (for newer programmers, "directory" is the old term for "folder"). ~ signifies the user's "home" directory.

If you are unfamiliar with Linux, there are several commands that are helpful to know:

`pwd`: This stands for "print working directory." This tells you what directory you are currently working in. If you do this when you first log in, it should say `/root`. Linux directories, rather than starting with a drive letter, start with a slash (`/`) as the top-level directory. `/root` is the home directory for the root user.

`mkdir`: This stands for "make directory." This creates a new directory inside the current directory.

`cd`: This stands for "change directory." If you give it a directory name, it will go to that directory. If you type in the command without any parameters, it will take you to your home directory. The command `cd /` will take you to the root directory (note that the root directory is a term for the top-level directory, not the root user's directory). If a directory name starts with a slash, the `cd` command will assume that it is an absolute path, starting with the root directory. If a directory name starts with a tilde (`~`), the `cd` command will interpret the path relative to your home directory. Otherwise, it will interpret the path as being relative to the current directory.

`ls`: This stands for "list," which gives you a list of files in the current directory. To see the file permissions, add the option `-l` to the command. To see hidden files as well, add the option `-a`. Therefore, to see *both* hidden files and file permissions, type in `ls -l -a`.

nano: Nano is your easy-to-use text editor. If you are going to run machines as part of your job, you should also learn vim as it will be more efficient for you, but that is a more difficult task. Nano is easy to use and will be sufficient for getting started. If you want to create a file in the current directory called test.txt, type in nano test.txt and start typing. The key combination control-o will save (i.e., output) your file, and control-x will exit.

systemctl: This command handles starting and stopping system services on certain Linux distributions, including CentOS. You will be introduced to how to use it later in this chapter.

logout: This exits the current user session. You can also do this by typing exit or control-d.

I encourage you to take some time playing around with the commands mkdir, cd, ls, and nano. Try creating a new directory, going into it, and creating a new file there. Then, try logging out, logging back in with ssh, finding your file, and viewing it. Do this several times until you are completely comfortable with the process of logging in, logging out, navigating directories, and editing files.

After you are comfortable making files and directories inside your home directory, you should branch out and look around at other directories. You shouldn't edit files there yet (files outside your home directory may mean something important to the operating system), but there is no harm in looking around.

To get started looking around, go to the root directory (cd /) and look around (ls). You will see a number of directories, most of which are documented in the Linux Filesystem Hierarchy Standard (see www.pathname.com/fhs). Even though the Filesystem Hierarchy Standard is a good place for general information about what the directories are for, it is no longer strictly adhered to, so don't be surprised by some amount of deviation. However, in short, `/etc` contains server configuration information, `/home` contains home directories for users other than root, `/usr` contains installed programs, `/opt` contains customized programs and other server-specific items, and `/var` contains information that changes regularly (e.g., log files, caches, queues, etc.).

3.3 Updating Your System

The first thing you should do after your system is up and running is to update the server with the latest upgrades and security packages. CentOS uses `yum` to manage system software installations. `yum` handles downloading, installing, upgrading, removing, and verifying software packages simply and safely. When you ask `yum` to install or update a package, it is smart enough to find the package on a remote server, verify the package's authenticity, look up and install any other software that the package depends on, and keep track of all installed packages and their files. `yum` can only be run by the root user, but so far that is the only user we have available.

The first thing you should do when you have a new server is to update all of its installed packages to their latest versions. Fortunately, this is really easy with `yum`. Just run the following command:

```
yum -y update
```

This will often download a *very large* number of packages—that's just fine. CentOS is continually fixing bugs and solving security issues in every piece of software in the distribution, and so these updates can get large. However, CentOS is also very careful to make sure that the revisions that it includes do not include any incompatible upgrades. So, by running `yum update`, you are keeping yourself up to date, and you are unlikely to accidentally break anything by running it.

3.4 Running the Web Server

By default, the Linux distributions that come with Linode only install the absolute essentials. This is actually great, because one of the primary means of maintaining security is to install only what you absolutely need, which minimizes the number of potential vulnerabilities that you are exposed to. However, by default the web server is not installed.

If you put in the IP address of your web server into your browser, your browser will respond that it cannot connect to the computer. This is because the web server is not running yet. To install the web server, we will use CentOS's yum package manager. This book covers the Apache web server, known to the computer as httpd, but there are other possibilities as well.

To install httpd, run:

```
yum -y install httpd
```

It doesn't matter what directory you are in when you run yum. It will install the packages to the correct directory no matter where you run it from. yum will list the package you want to install, as well as the other packages required to run it.

Now your web server is installed, but it is not *running*. To run the web server, simply type in:

```
systemctl start httpd
```

This command also does not care what directory you are in.

Now your web server is running, but you probably still can't connect to it. This is because CentOS comes with a firewall running by default. Therefore, we will need to add holes in the firewall to allow access to the web server from outside.

To do this, issue the following commands:

```
firewall-cmd --add-service http
firewall-cmd --add-service http -permanent
firewall-cmd --add-service https
firewall-cmd --add-service https --permanent
```

These commands will add both HTTP and HTTPS to the list of services that remote users can connect to. `firewall-cmd` manages your firewall. Adding a service (`--add-service`) allows access to that service. Running it without the `--permanent` flag modifies the current firewall. Adding the `--permanent` flag tells the firewall to have that rule in place when the server is rebooted. To enable the rule now and have it still enabled if the server is rebooted, you need both commands.

You can see the list of allowed services by issuing the following command:

```
firewall-cmd --list-services
```

When these commands have been run, you can simply go to your server's IP address in your browser, and you should get a test screen like Figure 3-5.

This means that your web server is up and running like it should be—congratulations!

However, there is one more issue to consider. Even though the server is running right now, if you reboot your machine, it won't be running when it starts. To make sure this service runs at startup as well, issue the command:

```
systemctl enable httpd
```

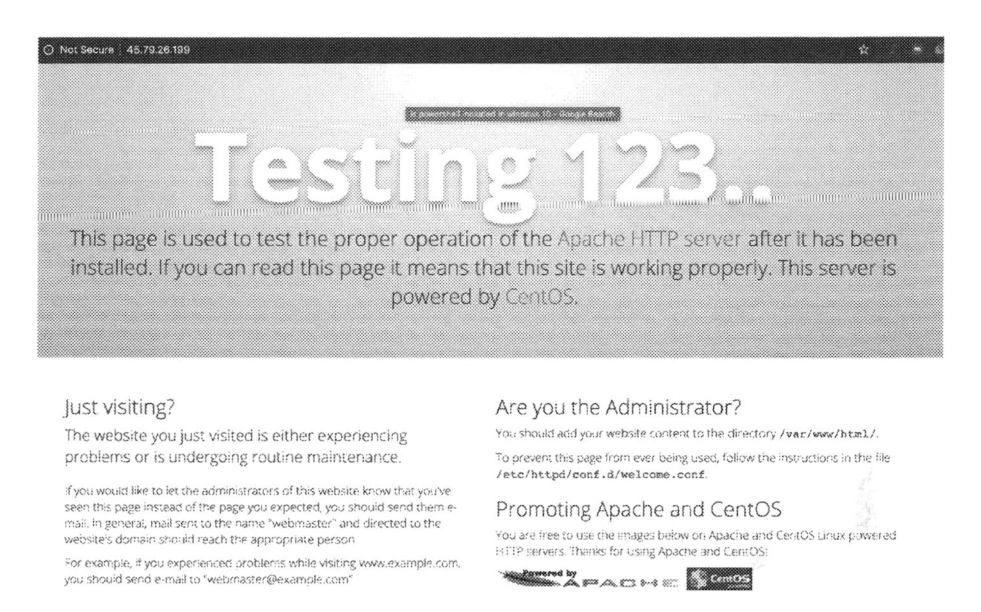

Figure 3-5. *The Web Server's Test Screen*

To test it out, go to your node's dashboard, and in the top-right hand of the page, it should say "Running." If you click that, it will show you a "Reboot" button. Click that button to reboot your machine. While your service is rebooting, the test web page will probably go away at some point. However, once the progress bar for the reboot finishes, the test web page should be available again.

Rebooting will log you off since the computer is no longer running. However, you can just log right back in again, and you will be back in root's home directory.

3.5 Putting Up Your Own Web Pages

The test page we get is automatically generated if there is no content for the web site. To create content for the web site, you simply have to put some in the right place.

Go into /var (type cd /var) and look around (type ls). One of the directories you see will be www. This is the default directory for data served from the web server (i.e., web pages). Go into www (type cd www) and look around (type ls). The html directory is where you will put your HTML and PHP files. Go into that directory (type cd html). To verify you are in the right place, type pwd and it should tell you that you are in /var/www/html.

Now that you are in /var/www/html, you will create pages for the web service to serve up. Create the file index.html using nano index.html and put something in it (if you don't know what to type, just type in hello there or something similar). Use control-o to save the file and control-x to exit the editor. As soon as you have created the file, you can go to your IP address, and that file will show up as the default page.

3.6 Installing PHP 7

Since this is a book on web *application* development, we want to do more than just web pages. We need to enable scripting on the server side. Therefore, we need to install the application framework we are developing with, as well as the plugins to get it running with Apache. This book focuses on PHP 7. Unfortunately, CentOS 7 only has PHP 5 available. Therefore, we are going to have to load PHP 7 from another repository.

Two commonly used package repositories are the EPEL (Extra Packages for Enterprise Linux) repository and Remi Collet's repositories. It is easy to load new repositories for yum to find. Each repository has a URL that yum can load so that yum can use it for future installation commands. To enable these repositories, just type the following (the indented lines should be put on the same line as the previous line):

```
yum install -y
  https://dl.fedoraproject.org/pub/epel/epel-release-latest-7.
  noarch.rpm
yum install -y
  https://rpms.remirepo.net/enterprise/remi-release-7.rpm
```

Now we can install PHP 7. To do this, just type:

```
yum install -y php74
```

This installs the base PHP 7 package and its dependencies, but without much else. Once installed, while still in the /var/www/html directory, run the command nano test.php and enter the following script:

```
<?php echo "This is a Test"; ?>
```

After doing this, save the file with control-o and exit with control-x. You can now run the file directly using the command line. Type the command:

```
php74 test.php
```

It will run the code in your file and output the string This is a test just like the code says.

Now, open your web browser and go to http://MY.IP.ADDRESS.HERE/test.php. Notice that it did *not* run it as a PHP script. That is because Apache and PHP are not connected together.

We now need to *connect* PHP to Apache. This is done through the FastCGI Process Manager (FastCGI is the protocol that allows PHP and Apache to communicate). You can install this with the command:

```
yum install -y php74-php-fpm
```

This is a separate process, so it has to be enabled and started as well.

```
systemctl enable php74-php-fpm
systemctl start php74-php-fpm
```

Now, we have to configure Apache to connect .php files to the PHP 7 interpreter. Create a file called /etc/httpd/conf.d/php.conf using nano, and put in this text (the indented lines should go on the *same line* as the previous line).

```
LoadModule proxy_module modules/mod_proxy.so
LoadModule proxy_fcgi_module
   modules/mod_proxy_fcgi.so
ProxyPassMatch ^/(.*\.php(/.*)?)$
   fcgi://127.0.0.1:9000/var/www/html/$1
DirectoryIndex /index.php index.php
```

Figure 3-6. *PHP Configuration File*

This code tells Apache to forward (called proxying) all requests for files ending in `.php` to the FastCGI service we previously installed. The last line tells Apache that it can treat `index.php` as a directory index, meaning that if someone leaves off a filename, and an `index.php` file is available, it can serve that file up as the default page in that directory.

Now, to get it to all work, restart Apache:

```
systemctl restart httpd
```

Now you should be able to hit the file with your browser, and it should be processing through PHP 7. Hooray!

While PHP is installed, you only have the base package installed. To see all of the extensions you can install, run the command `yum search php74`. This will display a list of all of the available PHP packages, each of which can be installed with `yum install -y`.

3.7 Turning Off SELinux

SELinux is a security enhancement feature for Linux. While, in theory, SELinux can do a lot to minimize security risks on servers, in practice it is too unwieldy for practical use. SELinux is generally not needed for well-built sites and not enough protection for poorly built ones. Instead, it winds up adding a large system administration headache for very little gain. If you run SELinux, the most likely outcome is that you will spend days trying to figure out why something isn't working, only to find out that SELinux is preventing some basic operation for no good reason.

To turn off SELinux, edit the file `/etc/selinux/config` and change the line that says `SELINUX=enforcing` to say `SELINUX=permissive`.

Reboot your machine. When it comes back up, log in and issue the command `getenforce`. It should say `Permissive`. You are now all set to go.

Note that if you don't disable SELinux, the application built in this book will not work, because SELinux will prevent the application from connecting to the database.

3.8 Setting Up a User for Development

So far we have used the root user (i.e., the superuser). While there are many things for which the root user is required, normally you should spend as little time being the root user as possible for safety reasons. Since the root user is allowed to do anything, it is easy to violate safeguards as the root user to either damage the system or accidentally allow unauthorized entities access to your system.

Therefore, we will create a non-administrative user to do most of our tasks. This user will be named `fred`. To create the user, type the following:

```
useradd fred
```

This sets up Fred as a user on the system, creates a home directory for him, and gives him a user and group number. Now we need to set a password for Fred by typing in the following:

```
passwd fred
```

This will prompt you for a password and make sure that you typed it in correctly by asking you to repeat it. We want Fred to be able to add and modify files in the /var/www/html directory, so we need to grant him the ownership of those files. We do this with the chown command:

```
chown -R fred /var/www/html
```

This tells to change the owner of the /var/www/html directory and all of the files under it to fred. We can now log out or we can open a new window and ssh back into the machine as fred:

```
ssh fred@MY.IP.ADDRESS.HERE
```

This will land us in Fred's home directory. We can now cd into /var/www/html and modify the files as before, since they are now owned by Fred. Note that root, since it is the superuser, can still modify the files as well even though they are owned by Fred.

3.9 Transmitting Files to the Server

Now, most people don't like programming directly on their servers. Usually, they want to write the program on their own machine and then transmit it to the server. To do this, you will need a tool that supports the SFTP protocol. SFTP is basically FTP over SSH.

The easiest SFTP solution that works across Windows, Macintosh, and Linux is FileZilla (www.filezilla-project.org). You can use either the regular (free) or the professional (paid) version. To use FileZilla, after you install it, just open it up, click "File," and then click "Site Manager." Click the "New Site" button.

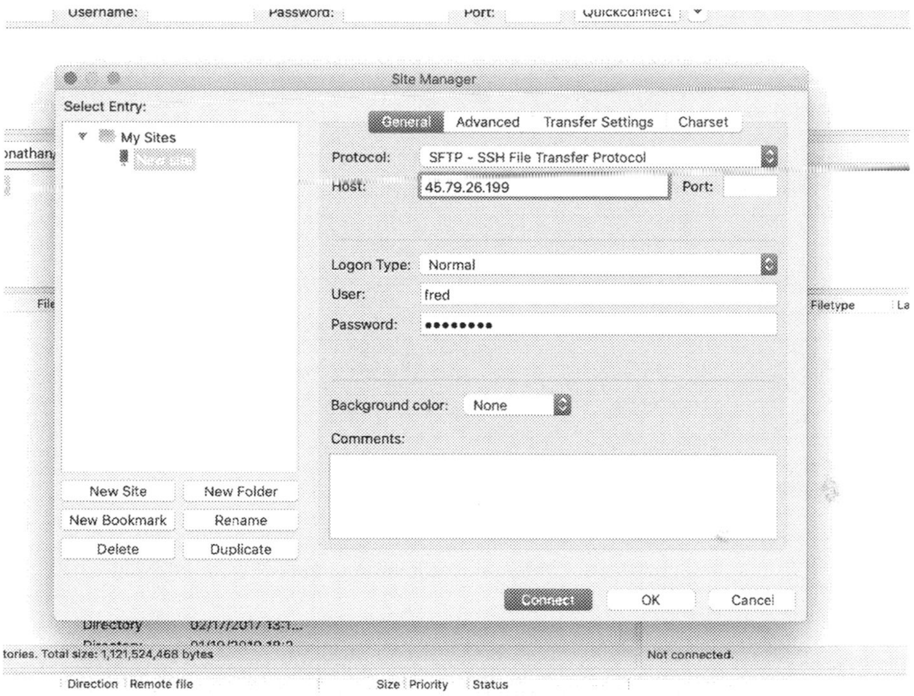

Figure 3-7. *Setting Up FileZilla to Connect*

Fill out this screen similar to Figure 3-7. In the "Host" box, put the IP address of your Linode server. In the "Protocol" box, select SFTP. Set the "Logon Type" to normal, and put the username as `fred` and the password as whatever you set for Fred's password. You can also change the name of the site to something memorable (I called mine "Linode Virtual Server").

When you are done, click the "Connect" button.

The first time you connect, it may bring up a dialog that says, "Unknown host key." This is fine—it is the same idea as when you first connected via `ssh`. The software just has never seen the server before. Click the box that says, "Always trust this host," and then click "OK." Once it is connected, it will look similar to Figure 3-8.

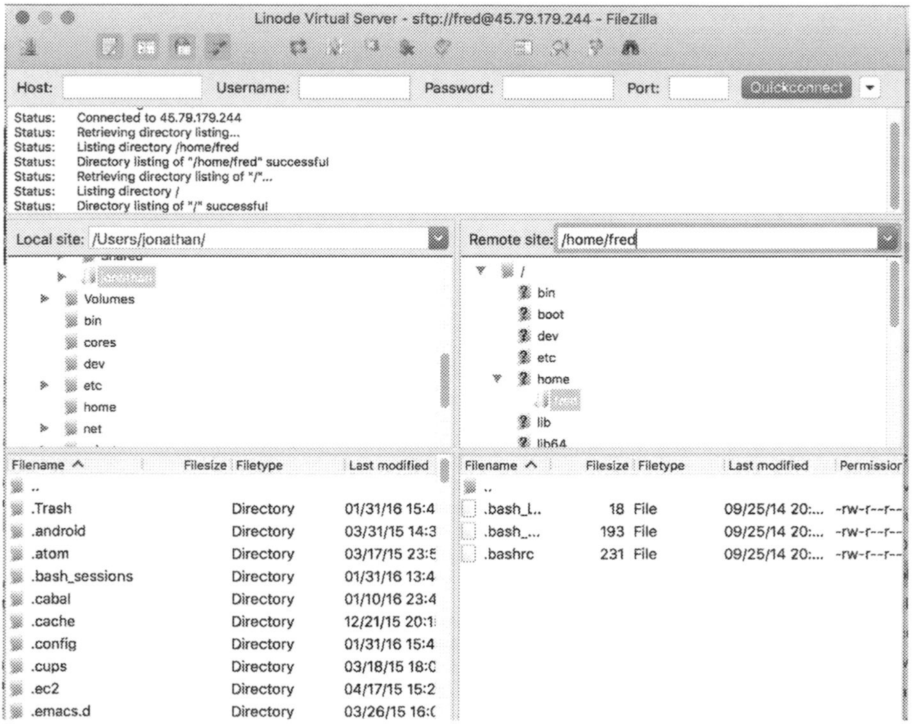

Figure 3-8. *FileZilla Connected to Your Linode Server*

FileZilla operates with two panes—the left pane is your local computer, and the right pane is the remote computer. It tells you which directory it is looking at in each one. What you need to do is set the local directory to wherever your files are you want to transfer, and the remote directory to wherever you want to put those files (presumably `/var/www/html`). Once these directories are set, the two panes below those list the actual files, and you can simply drag and drop them back and forth.

As an exercise, create a few simple PHP files on your local computer and transfer them to the server, and verify that you can see them through your web browser.

☞ EDITING PHP.INI

For some applications, you may need to modify the `php.ini` file. The version of PHP 7 we have installed puts the `php.ini` file in the directory `/etc/opt/remi/php74/`. However, only root can access this file.

If you need to transfer it with FileZilla, you will need to reconnect as root, with root's password. You can also log in via `ssh` and modify it directly with nano. In any case, always exercise care when modifying this file. Also, after modifying the file, be sure to restart the PHP process using

```
systemctl restart php74-php-fpm
```

This may seem to be a lot of work to get set up. While there are faster ways to get started, this method gives you several advantages. First of all, you are now more familiar than most people with how all of these pieces connect together. Second, you have a web server running PHP 7, and not some decade-old version. Finally, you have everything you need, and nothing that you don't, which will help you with keeping your web server secure.

☞ OTHER TOOLS TO INSTALL

Personally, I like my nodes packed with tools. I have been a Linux user longer than many of my readers have been alive, so I am pretty familiar with a large swath of tools. Thankfully, they are all easy to install (you must be logged in as root to do so).

In any case, the following are the install commands for the tools that I frequently use:

```
yum install -y git
yum install -y screen
yum install -y telnet
```

```
yum install -y bind-utils
yum install -y traceroute
yum install -y nmap
yum install -y strace
yum install -y perl
```

This book is not an introduction to these commands, but they are worth investigating.

CHAPTER 4

Creating a Simple Web App

In this chapter, we are going to be creating an extremely simple web application to use for demonstration purposes in subsequent chapters. The goal here is to get a full, end-to-end application up and running.

The application we are going to develop will simply be a guestbook, so that anyone can come and post a message into the guestbook.

4.1 Setting Up the Database Service

Any good web application has a database. My database of choice has always been PostgreSQL (`www.postgresql.org`). There is a myth that PostgreSQL is slow. There was some truth to that—in the 1990s.

However, starting with PostgreSQL 7, PostgreSQL has been a top performer, and it just gets better with every release. Additionally, PostgreSQL has *always* been fantastic with complex queries, and it remains so today. PostgreSQL aims at no-limit programming. For instance, in a PostgreSQL text column, you can store up to 4 gigabytes in a single column of a single row—and still sort by it. On many databases, most of your time is spent making the data match the preferred architecture of the database. I have found that, with PostgreSQL, the database is much more often already ready for your own data architecture.

J. Bartlett, *Building Scalable PHP Web Applications Using the Cloud*,
https://doi.org/10.1007/978-1-4842-5212-3_4

While this is not a book on PostgreSQL, we will discuss a few of its features related to clusters of nodes.

To install PostgreSQL, simply do the following as root (all of this section should be performed as the root user):

```
yum install -y postgresql-server
```

This will install all of the needed packages for PostgreSQL. To set up the initial database, type in the following:

```
postgresql-setup initdb
```

This creates all of the necessary directories and files for PostgreSQL to run. Next, we need to set up the authentication method for connecting to our PostgreSQL databases. PostgreSQL stores both its data and its configuration in the directory `/var/lib/pgsql/data`. The file which controls access to the database is `pg_hba.conf`.

Edit that file (type `nano /var/lib/pgsql/data/pg_hba.conf`) to add the following two lines to the top:

```
local all all trust
host all all all md5
```

The first line says to trust all connections coming in locally (i.e., not through the network). Therefore, we won't need a password when dealing with the database directly on the command line. The second line says that anyone can connect to the database over the network using an appropriate password. This would be somewhat unsafe (we don't want just anyone being able to connect to our database), except that by default the database only listens on the local address, 127.0.0.1, so right now you can't connect to it from outside anyway. Be sure to save the file and then exit the editor.

Note that even with the restrictions we have in place (local-only connections, firewalls, etc.), many people would consider the preceding configuration too exposed for their liking. The measures here are for balancing security and ease of learning. For more information about securing PostgreSQL, you should read the documentation on `www.postgresql.org` about the `pg_hba.conf` file.

Now it is time to turn on our database. To do that, enter the following:

```
systemctl enable postgresql
systemctl start postgresql
```

Your database system is now up and running. You have created the database *system*, but not the database itself. First, however, we need a database user. The `createuser` command creates a new database user:

```
createuser -U postgres -d -P gbuser
```

The command runs as the database admin user (`-U postgres`) and creates a new user named `gbuser` who can create databases (`-d`) and prompts you to set a password for this new database user (`-P`). When it prompts you, set the password to whatever you want and write it down so you have it later. We will use the password `mypassword` where needed in this book, but note that this would be a terrible password to actually use in production. You will not need to use the password when using the command line since we have it set to `trust`, but you will need it when connecting from your application.

To create a database as this user, type:

```
createdb -U gbuser guestbookapp
```

This creates a new database as the given database user. Now, to create the tables, we will need to log in to the database. The `psql` command will give you an interactive SQL session to your database. To use it, just type:

```
psql -U gbuser guestbookapp
```

The command prompt will switch to something like `guestbookapp=>`, which indicates you are in the database. To quit at any time, you can type `\q`. Like many of the system administration commands, PostgreSQL doesn't really care where in the filesystem you are when you run its commands. It is communicating to a database service, which is running in its own directory.

Now that we are connected to the database, we will create a single table using this command:

```
create table gb_entries(id serial primary key, name text, email
text, message text, created_at timestamp,
has_img bool default false);
```

The `id` field was created with type `serial`, which is PostgreSQL's mechanism for autonumbering. To view the table you just created, type `\d gb_entries`. When you are done, exit out of the database by entering in `\q`.

4.2 The PHP Code

Before we write the PHP code, we need to install some additional PHP libraries so that we can connect to our database. Install them with:

```
yum install -y php74-php-pgsql php74-php-pdo
systemctl restart php74-php-fpm
```

Our application will be simple:

- One file to hold the configuration information and common functions
- One file to show a list of guestbook entries
- One file to show an individual guestbook entry
- One file to enter a new guestbook entry
- A CSS stylesheet

This book presumes that you know a modicum of PHP and SQL, but even if you don't, the code should be straightforward enough to follow no matter what language you are familiar with. Therefore, the files will be presented here without too much comment. You will find the figures containing the code at the end of the chapter.

Figure 4-1 shows the common functions that are included by the other files. It has two functions for getting database connections—one for getting a read-only connection and one for getting a read/write connection. At this point, they both return the same connection (and they are indeed both read/write), but as we develop the application further, we will see how a lot can be gained by separating out connections that are used for reads only and those that are used for reading and writing. These functions both simply use PDO (PHP Data Objects) to grab a connection to the database using a connection string. Note that these connection strings contain the password to the database. Be sure to change the password on *both* of these connection strings to whatever password you entered for `gbuser` database user earlier.

It also has the `getHeader()` and `getFooter()` functions so we don't have to write as much HTML. Also, `h()` is used as a shorter version of `htmlspecialchars()` so that we can have more secure output.

Figure 4-2 is the PHP script that lists all of the entries in the database. This code simply creates a database statement, executes it, and iterates through the results.

Figure 4-3 shows getting a single entry out of the database. Again, there is a single SQL statement that is prepared and executed, and the results are displayed to the screen.

Figure 4-4 simply shows a form that will be used to create a new guestbook entry. This form posts its data to the program in Figure 4-5. That program creates a new record based on the values entered in. Then, after performing the SQL insert statement, it redirects the user back to the listing screen.

Finally, Figure 4-6 is a static CSS file that provides a small amount of style to the whole procedure. As mentioned in Chapter 1, if you don't want to type in all of these files yourself, you can get them all from GitHub at:

```
https://github.com/johnnyb/cloud-example-application
```

To use git directly on the server, you will need to install it as the root user with:

```
yum install -y git
```

After typing in all of the programs, send them up to the `/var/www/html` folder on your new server. Then, navigate your browser to `http://MY.IP.ADDRESS.HERE/list.php` and see if your program works! If it does not, you can check the error log for PHP-FPM by logging in as root and looking at the log file with the following command:

```
tail -200 /var/opt/remi/php74/log/php-fpm/error.log
```

This will give you the last 200 lines of PHP's error log.

Another log that will give you good information can be accessed by issuing:

```
tail -200 /var/opt/remi/php74/log/php-fpm/www-error.log
```

Fix any errors that you have and try again. The most likely error is that there was something mistyped in the program, or the password listed in the connection string doesn't match the password you set for your PostgreSQL user.

Another place to look for error messages is the web server's error log. You can look at the end of that log with the command:

```
tail -200 /etc/httpd/logs/error_log
```

Additional troubleshooting steps can be found in Appendix C if needed. If all goes well, you should have a screen that prompts you to create a new entry. Clicking the link will give you a form to fill out. Once you fill out that form, clicking "Submit" will add the new guestbook entry to the list. You can then click the individual entry to see the full information. If that is not what your application does, then something was probably entered incorrectly.

THE LIMITATIONS OF THE APPLICATION

The goal of this book is to get you up to speed on how to scale your applications. As such, other important aspects of development such as error handling, logging, sanitizing data, and security hardening are not covered. The goal is to convey an application that can be typed in quickly, is easy to understand fully, and does not require deep platform knowledge to follow or modify. Nonetheless, we have implemented some basic security practices, such as using `bindValue` to properly escape the values sent in via `$_GET` and `$_POST` and using `htmlspecialchars()` to escape them when sending them back to the user.

Good information about programming with security in mind can be found on the `www.owasp.org` web site.

```
<?php
  function getReadOnlyConnection() {
    return new PDO("pgsql:host=localhost;" .
      "port=5432;dbname=guestbookapp;" .
      "user=gbuser;password=mypassword"
    );
  }

  function getReadWriteConnection() {
    return new PDO("pgsql:host=localhost;" .
      "port=5432;dbname=guestbookapp;" .
      "user=gbuser;password=mypassword"
    );
  }

  function getHeader() {
?>
<html>
  <head>
    <title>Guestbook App</title>
    <link rel="stylesheet"
      href="guestbook.css" />
  </head>
  <body>
    <h1>Guestbook App</h1>
    <div id="main">
<?php
  }

  function getFooter() {
?>
    </div>
  </body>
</html>
<?php
  }

  function h($str) {
    return htmlspecialchars($str);
  }
?>
```

Figure 4-1. *Configuration and Common Functions (`common.php`)*

```
<?php
  include("common.php");
  $dbh = getReadOnlyConnection();
  $stmt = $dbh->prepare(
    "select * from gb_entries"
    " order by id"
  );
  $stmt->execute();
  $result = $stmt->fetchAll();
  getHeader();
?>
<a href="new.php">New Entry</a>
<ul>
<?php
  foreach($result as $row) {
    $url = "single.php?id=" . $row["id"];
?>
    <li>
      <a href="<?= $url ?>">
        Entry from
        <?= h($row["name"]) ?>
      </a>
    </li>
<?php
  }
?>
</ul>
<?php
  getFooter();
?>
```

Figure 4-2. *List All Guestbook Entries (`list.php`)*

```
<?php
  include("common.php");
  $dbh = getReadOnlyConnection();
  $stmt = $dbh->prepare(
    "select * from gb_entries" .
    " where id = :id"
  );
  $stmt->bindValue(
    ":id",
    $_GET["id"],
    PDO::PARAM_INT
  );
  $stmt->execute();
  $result = $stmt->fetchAll();
  $row = $result[0];
  getHeader();
?>
<p>
  <a href="list.php">Go Back</a>
</p>

<label>Name</label>
  <?= h($row["name"]) ?>
<label>Email</label>
  <?= h($row["email"]) ?>
<label>Message</label>
  <?= h($row["message"]) ?>
<?php
  getFooter();
?>
```

Figure 4-3. *Single Guestbook Entry (`single.php`)*

```
<?php
  include("common.php");
  getHeader();
?>
<h2>New Guestbook Entry</h2>
<form action="create.php" method="POST">
  <label>Name</label>
  <input type="text" name="name" />

  <label>Email</label>
  <input type="text" name="email" />

  <label>Message</label>
  <textarea name="message"></textarea>

  <input type="submit" />
</form>
<?php
  getFooter();
?>
```

Figure 4-4. *New Guestbook Entry (new.php)*

```
<?php
  include("common.php");
  $has_img = false;
  $dbh = getReadWriteConnection();
  $stmt = $dbh->prepare(
    "insert into gb_entries" .
    "(name, email, message, has_img)" .
    "values " .
    "(:name, :email, :message, :has_img)"
  );
  $stmt->bindValue(
    ":name",
    $_POST["name"]
  );
  $stmt->bindValue(
    ":email",
    $_POST["email"]
  );
  $stmt->bindValue(
    ":message",
    $_POST["message"]
  );
  $stmt->bindValue(
    ":has_img",
    $has_img,
    PDO::PARAM_BOOL
  );
  $stmt->execute();

  header("Location: list.php");
?>
```

Figure 4-5. *Create the Guestbook Entry (`create.php`)*

```
body {
  background-color: black;
  color: white;
}

div#main {
  padding: 1em;
  border: 1px solid black;
  background-color: white;
  color: black;
}

label {
  display: block;
  font-weight: bold;
}
```

Figure 4-6. *CSS File (guestbook.css)*

CHAPTER 5

Setting Up a Basic Cloud Cluster

At this point, we have a simple cloud application which works on a single server. While it is nice that we can put this on a server that we built just by pointing and clicking, this doesn't take full advantage of the cloud. One of the goals of the cloud is to create an application *cluster*—a set of servers that work together to solve the problem, where the computing power can be expanded and contracted as needed.

5.1 A Simple Two-Tier Architecture

In this chapter, we will explore a simple, two-tier architecture. This architecture will consist of

- A database server
- A set of web servers
- A load balancer that manages traffic among the web servers

J. Bartlett, *Building Scalable PHP Web Applications Using the Cloud*,
https://doi.org/10.1007/978-1-4842-5212-3_5

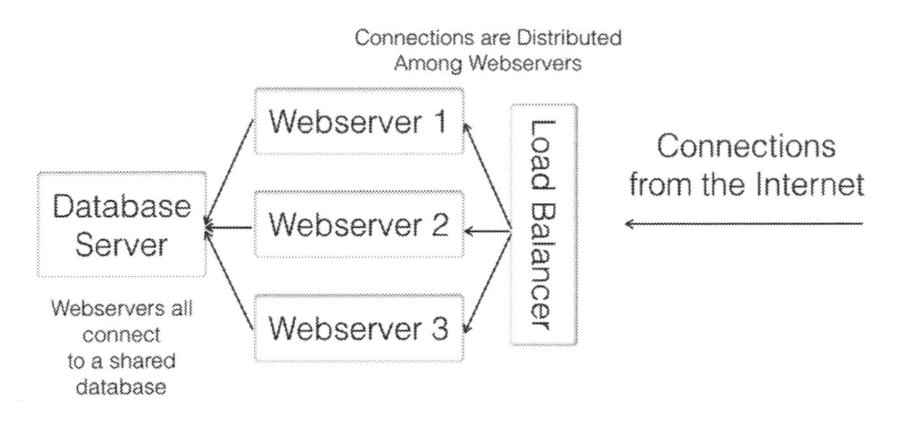

***Figure 5-1.** Diagram of a Simple Two-Tier Architecture*

The basic structure of the architecture is shown in Figure 5-1. All of the connections come in to a single load balancer, whose job it is to forward connections to one of several web servers. The load balancer not only forwards connections, but it also monitors the health of the individual web servers and will stop sending a web server connections if it stops responding. Each web server then shares a single database server.

In developing cloud applications, a programmer needs to know not only how to set up a cluster but also how to analyze it. If you look at the diagram in Figure 5-1, you can see that all of the web servers depend on a single database server. This makes the database server the *limiting factor* of the cluster. Nearly every cluster, no matter how well designed, has some limiting factors. The goal is to minimize their impact on your architecture.

Because this application architecture is limited by the single database node, it is best used for small- or medium-sized deployments where most of the processing is done on the web servers rather than on the database. The example application, because of its simplicity, actually does very little processing on the web server. Nonetheless, this chapter will show you how to set up the servers to deploy it in this configuration.

5.2 Replicating a Node

The basic two-tier application described in Figure 5-1 indicates that we will need several server nodes. We could accomplish this simply by booting up new machines with a blank copy of CentOS, and configure each one individually. However, since we have already spent time setting up the current server to work the way that we want it to, we should take advantage of the time we spent configuring everything.

Linode offers several services that can accomplish this task, each with their own benefits and drawbacks. Linode has the ability to create saved images which can be used to directly create new nodes (instead of choosing an operating system, you can choose your saved image). Linode also has a cloning service, which allows you to clone existing machines, if both the source and destination machines are turned off (this is not a hard and fast rule, but you can wind up with consistency problems if you try to clone a running machine). Finally, you can make clones from a machine's backups.

I prefer making clones from backups because (a) you should be backing up your servers regularly anyway, (b) you don't have to have a machine turned off and doing nothing while you create new servers, and (c) this forces you to work with the backup system and be comfortable with it while you are also learning to clone servers (someday you will need to restore from backup, so it is good to get comfortable with the process before you need it). The Linode image service *can* work for this, but it has too many limitations for real production use. In order to prevent users from using the image service as a backup service, they have limited the size and number of images, but most of my machines are usually heftier than the minimum size allowed for Linode images. Note that other cloud services (such as DigitalOcean) offer similar services but with different sets of restrictions.

If we want to boot up a new server that is an identical clone of the existing server, then all we have to do is back up the current server. What we will need to do is first enable backups on our server node. To do that, just log in to Linode, click your node in the list, and click the "Backups" tab off of your node's dashboard. This will give you a button that says, "Enable backups for this Linode." This adds a small monthly charge, but it is certainly worth it. Click the button, and your machine will automatically have weekly, daily, and ad hoc snapshots available. The screen should now look similar to Figure 5-2.

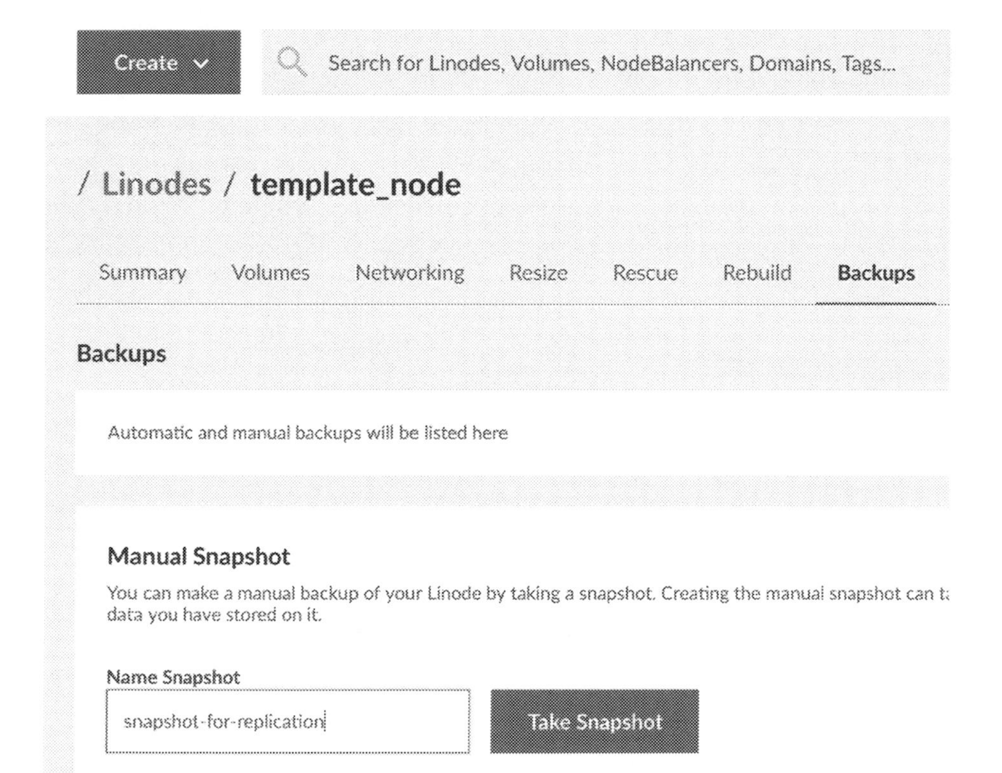

Figure 5-2. *The Linode Backup Management Screen*

What we want to do now is to take a snapshot of our machine. Put in a name for your backup (I called mine "snapshot-for-replication"), and click the button that says, "Take Snapshot." Because Linode servers are all on SSD drives, snapshots are very quick, usually only taking 5–10 minutes.

☞ OTHER BACKUP SETTINGS

Because backups do in fact slow down your disk somewhat, Linode gives you the ability to specify your backup window. You can choose both a time of day when you want to have your backups take place and the day of the week which will be counted as a "weekly backup." The time of day should be when your system is having the lightest usage, while the day of the week should be before any major batch processing. For instance, if you do batch processing on Fridays, setting the backup day to Thursday will make sure you have a "before" snapshot before the major processing occurs. If you don't do large-scale batch processing, the day of the weekly backup doesn't really matter.

Once the backup has started, we can go to our next task, which is creating a new server. The new server will be our database server.

To create the machine, go to "Create" then "Linode" as we did in Chapter 3. However, instead of choosing a distribution, under the "Images" heading, you want to choose "My Images" and then "Backups." This will show a list of your nodes which have backups available. Click your node, and then it will show the available backups, which will include the one we just created. See Figure 5-3 to see what this should look like.

You can create as large as a machine as you want, but just for practice, you might as well use their smallest machine size. This node needs to be in the same datacenter as your other one (otherwise they won't be able to privately, cheaply, and quickly talk to each other), but Linode will automatically put a new node created from backup into the same datacenter.

For the Linode label, let's use the name `dbmaster` so that we know that this node will be used for the master database. Having a bunch of nodes without names (or with bad names) quickly becomes difficult to manage, so be sure to *always* give your nodes descriptive names.

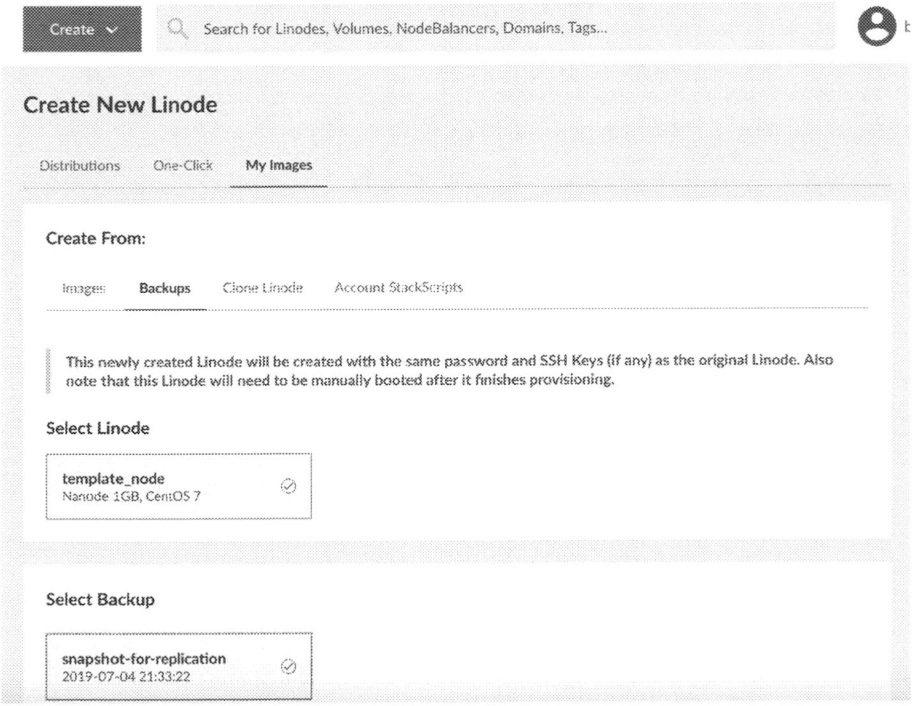

Figure 5-3. *Booting a New Linode from a Backup*

Now click "Create" to build your new node. When creating from a backup, you will have to boot the node yourself once it is fully created. On the top right of the screen, it should say "Offline." Click that, and choose "Power On" to boot your new machine. You can now `ssh` to this new machine using the same users and passwords that you set up for your initial machine. In fact, it should also have the same application running that you had created. It is an exact copy of the other machine, with only the network settings having been modified.

RESTORING TO A LARGER MACHINE

If you restore a backup to a larger machine, it may not utilize all of your purchased disk space. Since the partition was directly copied, the old partition will be the same size it was on the server it was copied from, which may be smaller than the space that you have available.

To solve this, first power down your server. Next, on your node's dashboard, click the "Advanced" tab. You can either add another disk to use the remaining space (this is harder to manage, so I don't recommend it) or resize your main disk to utilize all of your space. To resize the disk, find the main disk under "Disks." It should be named something like "CentOS 7 Disk" (not the disk labeled "Swap Image"). Click the ellipses (i.e., ...) next to your main disk and choose "Resize." You can then set the size to the maximum value it tells you, and then click the "Resize" button.

When it is done resizing, you can power your machine back on, and you are now all set.

5.3 Setting Up Your Private Network

Since we have two machines, we need them to communicate. They *could* communicate via their public IP address, but this leads to several problems. First of all, if you have services that you don't want to be available publicly (like your database), it is more difficult to prevent the public from getting access to them if you only have a public IP address. Additionally, Linode charges money for traffic on your public IP address, so, if you communicated via that IP address, Linode would charge you for *internal* traffic. Having a private IP address is important, therefore, because it lets the computers communicate with each other over a fast, free, more secure, internal network.

To get around these problems, Linode allows you to set up an internal network for your servers. All of the servers within an account share an internal network if they are set up for it. To add a server to your internal network, you just need to go into your node's "Networking" tab and click "Add Private IPv4." It will give you some additional information about private addresses, and you can just click "Allocate" to continue. This will assign a private IP address to the computer. Because all of the web servers will be communicating with this server, you will want to write down the private IP address that is generated. From here on out, we will refer to this address as `DB.MASTER.PRIVATE.IP`, so always replace this with the private IP address of the server that you just wrote down.

Note that on some cloud platforms (including Linode), the private IP addresses are not completely private. That is, other cloud customers in the same datacenter may be on that network. Therefore, while it is certainly safer on the private network than on the public network, private networks on clouds don't guarantee that only our own computers will be connecting. Therefore, on a production system, you will still need to take precautions to prevent unwanted access even on the internal network. However, Linode does filter traffic to each node, so you shouldn't have to worry about anyone snooping on data traffic on the internal network.

You will need to reboot the node to complete the process.

Once it is done booting, you can still see the application at `http://NEW.NODE.PUBLIC.IP/list.php`, but you will not be able to see it on the private IP address, because, as noted, it is private.

To see the list of addresses on the server, log in as root and issue the command:

```
ip addr show
```

This should print out all of the IP addresses assigned to this node.

☞ PRIVATE IP ADDRESSES

If you aren't familiar with IP addressing, some IPv4 addresses have been reserved for use in internal networks. These addresses include

- `192.168.X.X`
- `172.16-31.X.X`
- `10.X.X.X`

None of these IP addresses are allowed for public communication on the Internet.

Therefore, when setting up your internal network, Linode will choose IP addresses from these pools to configure your machine.

One other infamous IP address is `127.0.0.1`, which is known as the *loopback* address, which is an IP address that a machine can use to refer to itself (in fact, the whole range `127.X.X.X` is reserved for this purpose, but only `127.0.0.1` is normally used).

5.4 Handling Database Connections from Other Servers

This machine now has a full copy of the application and the database on it. However, it is still configured to be a single-server system. We need to configure it to be the master database for a cluster of machines. This section, which should be performed as the root user, will show what you need to do to make that happen.

Since this machine is a clone of `template_node`, that means that all of the users, programs, and configurations were copied over to this node. Therefore, you can `ssh` into the machine as root using the password you set previously.

To use this as the database server for the other nodes, you will need to enable the database to listen for connections from those nodes. By default, PostgreSQL only listens for connections on the localhost interface. What we *don't* want is for PostgreSQL to listen for connections on the public Internet. Therefore, we want to configure PostgreSQL so that it listens for connections on both localhost and its *private* IP address.

To do that, as the root user, open up the file `/var/lib/pgsql/data/postgresql.conf` with `nano`, and change the line that says `listen_addresses`. Change that line to read:

```
listen_addresses = 'localhost,DB.MASTER.PRIVATE.IP'
```

Be sure to replace `DB.MASTER.PRIVATE.IP` with the *actual* private IP address of your node. If there is a comment mark (#) at the beginning of the line, be sure to remove it; otherwise the command won't be active. Save the file with control-o (just hit the return key to verify the filename if it asks). Then, to exit, use control-x. Now restart PostgreSQL with the command:

```
systemctl restart postgresql
```

Additionally, you will want to open up the firewall so that it can accept remote connections for PostgreSQL. You can do this with:

```
firewall-cmd --add-service postgresql
firewall-cmd --add-service postgresql --permanent
```

Normally, I keep the web server running on the database server just so I can check it. However, if you wanted to, you can turn off the web server with the commands:

```
systemctl stop httpd
systemctl disable httpd
```

5.5 Setting Up a Web Server

Now that we have the database configured, it is time to set up our web servers.

Note that we will not actually use template_node as a server. I like to keep a small machine around that is simply used as a template for future boxes, especially for webnodes. This way, I can have one small machine that is up-to-date, backed up, and so on, and when I'm ready to create a new "image," I just create a named backup to use. Note that this will overwrite the previous snapshot image, but for my purposes, that is usually okay. If you need to keep old versions around, just keep a template node around for each configuration you want to maintain.

Now we will configure template_node for being a web server template. There are only three changes we need to make:

1. Turn off the database on this machine.
2. Change the code for the web application to point to our new database.
3. Enable a private IP address on the machine to enable it to use the private network.

To accomplish the first task, all we need to do is log in to the template_node machine as root and run:

```
systemctl stop postgresql
systemctl disable postgresql
```

To accomplish the second task, we need only modify the common.php file. I would suggest modifying it on your local machine and using SFTP to transmit the new file. However, you can also use nano on the server to modify it directly. All you have to do is to change the connection string. Where it currently says host=localhost, change it to read host=DB.MASTER.PRIVATE.IP, where DB.MASTER.PRIVATE.IP is the private IP address of your `dbmaster` node. You will need to make this change twice—once in the `getReadOnlyConnection()` function and once in the `getReadWriteConnection()` function. Once you are done, load the code back onto the server with SFTP.

At this point, the code will not work because it cannot access the `dbmaster` machine. This is because PostgreSQL is *only* listening on its private IP address, and `template_node` does not yet have a private IP address to communicate on. Therefore, you need to add a private IP address to the machine, so that it can connect to `dbmaster` on its private IP address.

Use the process outlined in Section 5.3 to create a private IP address for the machine (don't forget to reboot afterward!).

Once you have completed these steps, your `template_node` machine should be able to connect to `dbmaster`, so test it out. Go to the IP address of your `template_node` server and see if it can still function. If it does, then congratulations, because you just implemented a small, two-tier system!

Now, as I said earlier, we actually aren't going to use `template_node` to actually serve requests. The goal is to use it so that we can easily boot up new web server nodes to expand capacity as we need to.

Therefore, now that `template_node` is fully set up to be a web server, take a new backup snapshot. This step is critically important. Anytime you make a change in `template_node`, you should make a new backup snapshot so that new nodes you create from it will have your new changes (though it doesn't affect existing nodes at all).

We will now create three (or however many you want) web server nodes for our cluster.

Here are the steps for each new web server node:

1. Create a new node using the steps in Section 5.2. Be sure that (a) it gets created in the same datacenter as `dbmaster`, and (b) set the name of the node to `webnode1` (or 2, or 3).

2. Add a private IP address to the node so that it can connect to `dbmaster` on the private network using the steps in Section 5.3.

3. After the machine finishes booting up, verify that it is fully functional by looking at your web application on the public IP address of the node (i.e., `http://WEB.NODE.PUBLIC.IP/list.php`).

At the end of this, you should have three machines, `webnode1`, `webnode2`, and `webnode3`, each of which can act as a front end to your web application. Now you just need to link them all together, which will be covered in the next section.

5.6 Setting Up the Load Balancer

Now we have three front-end machines, all connected to a single database. How do we connect them together? One way we could do this is to set up a DNS round-robin scheme. The way this works is to set up *multiple* A records in DNS for a single hostname. Then the browsers themselves will pick which host they want to connect to. The problem with this is that if one of your machines goes down, there is no way to direct users away from that IP address. Linode actually has some support for this sort of failover now, but its usage is outside the scope of this book.

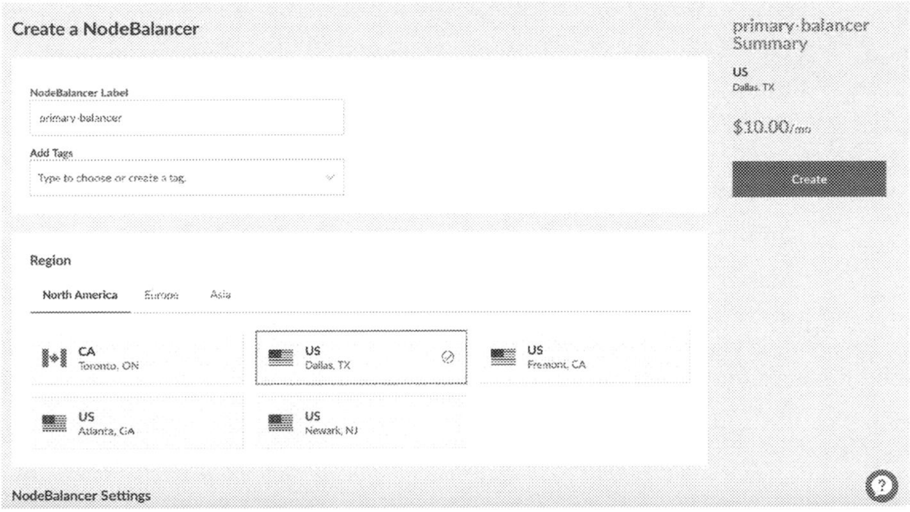

Figure 5-4. *Creating a Node Balancer*

Load balancers are a much more turnkey solution. A load balancer sits out in front of your cluster and takes connections for you and then forwards those connections to the servers that are available for handling them. Additionally, if one of your servers fails, the load balancer will detect this and move traffic onto the remaining servers. Then, when your server recovers, the load balancer will also detect this and move traffic back to the server.

Setting up a load balancer in Linode is easy. Linode calls its load balancers "Node Balancers." To set up a Node Balancer, click "Create" and then select the "Node Balancer" menu item. This will bring you to the screen shown in Figure 5-4.

Just like everything else, you need to

- Set the name of the balancer (we will use "primary-balancer").
- Put the balancer in the same datacenter as your nodes.

Additionally, you need to add some additional configuration, as shown in Figure 5-5. Be sure the following are selected:

- "Port" should be set to "80."
- "Protocol" should be set to "HTTP."
- "Algorithm" should be set to "Round Robin."
- "Session Stickiness" should be set to "None."
- "Active Health Checks" should be set to "None."
- "Passive Health Checks" should be on.

After that, you need to add at least one node to your balancer (more can be added after creating the balancer). Put in the name of the node in the "Label" field (we'll assume `webnode1`). Then, in the "IP Address" field, choose the IP address for your node from the drop-down. Make sure the port is set to "80."

When these settings are completed, click "Create" to create the balancer.

Once your balancer is created, you can add the rest of the nodes by going into "Configurations," then "Port 80," and then "Add Node" at the bottom. Add as many nodes as you created. Then click "Save" when done.

It sometimes takes a few minutes for the balancer to add a node to its list. To check on the status, go back to the node balancer configuration screen. Each server will have a "Status" field next to it. When the status is "Up," the server is successfully connected to the balancer.

If you wanted to have a load balancer for another port, you can use the "Add Another Configuration" button for this.

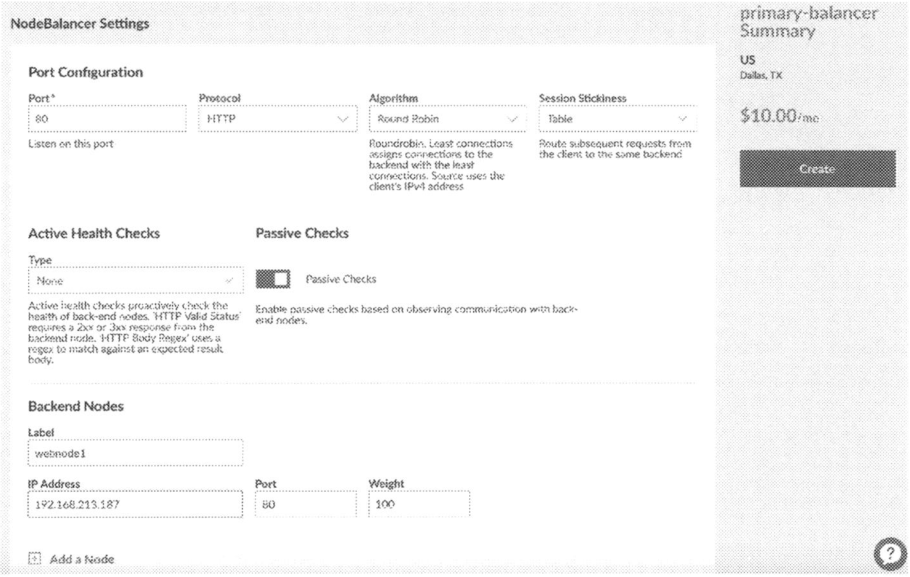

Figure 5-5. *Node Balancer Additional Settings*

☞ OTHER NODE BALANCER OPTIONS

The node balancer has a lot of options available. Here is a description of some of the important ones.

Port: This is the TCP port that the node balancer should forward. This will normally be 80 (HTTP) or 443 (HTTPS). We will use port 80 for our examples.

Protocol: This is how you want the server to handle forwarding your requests. If the protocol is set to TCP, then the *only* thing that the balancer does is forward the connection on to you. If it is set to HTTP or HTTPS, then the server will actually handle some parts of the connection for you. HTTP is what we will use for this book. HTTPS adds an additional boost for secure sites, because the load balancer will handle the SSL connection for you, thus removing a significant chunk of the load from the server (you also upload the certificate and key information to the load balancer for processing). When doing HTTPS, you probably want the port on the balancer to be different than the port on the machine. In the case of HTTPS, you should set the balancer to connect to the unencrypted port 80 on the servers. If you wanted to do HTTPS that was handled by your machine instead of the balancer, you would just choose TCP (instead of HTTPS) as the protocol. Because of the added complications of obtaining and installing certificates, for the examples in this book, choose port 80 and HTTP for load balancing.

Algorithm: This is the way that the load balancer will determine which server to forward it to. Round Robin is the default and should work fine.

Session Stickiness: This is whether or not a given user should continue to connect to the *same* web server once it has made an initial connection. This is important *only if* you store session information on your web servers. Imagine if your web server has important session information, but the next request goes to a different server! Therefore, this option allows you to configure whether and how clients get matched to servers. Our application doesn't have session information, so it should be set to "None." If your application uses local session information (or local caching, as we will see in Section 6.1), I would choose the HTTP Cookie method, as it won't get in the way of your load tests like the "Table" method will. Since all requests come from the same IP address, the "Table" method will simply direct your entire load test to a single server rather than spreading it out and make it look like your load balancing isn't giving you any help.

Active Health Checks: This allows you to specify a URL for the load balancer to hit in order to check the status of the web server. You can configure the balancer to either just check for an HTTP status or also look for a specific string in the response body.

Backend Node: Weight: This sets the preference of the node balancer for this server. A higher weight gives the server more connections.

Backend Node: Mode: This sets the mode of the node in the balancer. "Accept" is for normal operation. "Reject" essentially turns off this node, meaning that the balancer will not send this node any more requests. However, if you have session stickiness turned on, you might not want to move directly from "Accept" to "Reject." "Drain" tells the balancer to *only* accept connections from clients which have sessions on this node. "Backup" says to accept connections only if all other nodes are down.

Once you have all of the webnodes added to your node balancer, you can now view your cluster by going through the IP address of the node balancer itself, and the node balancer will forward your requests to one of the machines in the cluster. You can find the IP address of your node balancer by clicking "Node Balancers" in the main menu. The IP address will be listed next to your balancer in the list. It is also listed on the right-hand side of the node balancer's "Summary" screen.

5.7 Measuring Scalability

The simple web application we have developed does not benefit too much from the architecture presented here. Our web application is pretty much just a simple shell for a few database queries. Therefore, simply adding front-end boxes won't help the fact that our application is database limited. Separating the database from the web servers will give us some boost, as it allows the database server to focus on only database connections. However, at the core, everything we do in this app is just a thin wrapper around database queries.

But how do we *measure* the amount of scalability our app has? One common, simple tool for measuring the scalability of an app is ApacheBench. ApacheBench is standard on Macintosh and most Linux distributions. For our case, we can actually run ApacheBench from our `template_node` server to test the rest of our cluster.

To do this, log in to your `template_node` server as either `root` or `fred`. To run a simple ApacheBench session, just type:

```
ab http://BALANCER.IP.ADDRESS.HERE/list.php
```

Obviously, replace `BALANCER.IP.ADDRESS.HERE` with the IP address of your load balancer.

This will send a single request to the load balancer and document the amount of time that it took to process. The output will look something like Figure 5-6.

Since this only benchmarked a single request, there is not much interesting information in it. It says that the request took 9.674ms (milliseconds), and, extrapolating that out, it estimates that we can serve up to 103.37 requests per second.

Now, ApacheBench has several options that allow us to exercise servers more fully. The `-n` option tells ApacheBench how many requests to make (we did 1 by default). The `-c` option tells ApacheBench how many concurrent (i.e., simultaneous) connections to make. Without `-c`, ApacheBench will just run one request at a time. However, if you add `-c 50`, ApacheBench will always keep 50 active requests with the web server.

Therefore, to exercise the cluster, I did the following:

```
ab -c 50 -n 1000 http://BALANCER.IP.ADDRESS.HERE/list.php
```

This sends a total of 1,000 requests, making sure there are always 50 active at a time. This gave the results shown in Figure 5-7.

This says that with 50 concurrent requests, our *average* time per request falls to 5.136ms, but the time for *each* request grows to 256.822ms. This is not really a problem, as the average time per request is the most important for capacity planning. We are also told that, at this pace, our servers can handle up to 194.69 requests per second.

```
Server Software:          Apache/2.4.6
Server Hostname:          BALANCER.IP.ADDRESS.HERE
Server Port:              80

Document Path:            /list.php
Document Length:          383 bytes

Concurrency Level:        1
Time taken for tests:     0.011 seconds
Complete requests:        1
Failed requests:          0
Write errors:             0
Total transferred:        560 bytes
HTML transferred:         383 bytes
Requests per second:      88.63 [#/sec] (mean)
Time per request:         11.283 [ms] (mean)
Time per request:         11.283 [ms] (mean,
   across all concurrent requests)
Transfer rate:            48.47 [Kbytes/sec]
   received

Connection Times (ms)
              min  mean[+/-sd] median   max
Connect:        1    1   0.0      1       1
Processing:    11   11   0.0     11      11
Waiting:       10   10   0.0     10      10
Total:         11   11   0.0     11      11
```

Figure 5-6. *Example Output of ApacheBench for a Single Request*

```
Server Software:          Apache/2.4.6
Server Hostname:          BALANCER.IP.ADDRESS.HERE
Server Port:              80

Document Path:            /list.php
Document Length:          383 bytes

Concurrency Level:        50
Time taken for tests:     4.057 seconds
Complete requests:        1000
Failed requests:          0
Write errors:             0
Total transferred:        560000 bytes
HTML transferred:         383000 bytes
Requests per second:      246.47 [#/sec] (mean)
Time per request:         202.865 [ms] (mean)
Time per request:         4.057 [ms] (mean,
   across all concurrent requests)
Transfer rate:            134.79 [Kbytes/sec]
   received

Connection Times (ms)
              min  mean[+/-sd] median   max
Connect:        0    1  31.6      0    1001
Processing:    42  193  59.4    181    1217
Waiting:       40  193  59.4    181    1217
Total:         43  194  67.1    182    1218
```

Figure 5-7. *ApacheBench Results for 1,000 Requests*

This sounds great, except that when you measure it against a single machine (which you can do by replacing the IP address with one of the IP addresses of your machines), you get close to the same results. The load balancer can handle a little higher pressure when the number of concurrent requests skyrockets, but overall they both essentially have the same results. This is because our app is almost entirely just a shell for the database. Therefore, this number reflects the maximum capacity of our database server. To see that this is true, you can temporarily resize your database server to a larger machine.

To resize your `dbmaster` server, go to the `dbmaster`'s dashboard, click the "Resize" tab, and choose a new plan (I chose Linode 4GB). Now click "Resize this Linode Now." After a few minutes of downtime, your Linode will be resized, and it will keep its IP address! Once it is done resizing, it will boot up and you will now have a resized server!

You can now run ApacheBench on this configuration and see that resizing the database server gives you a huge advantage over your previous configuration. After doing this experiment, I have resized `dbmaster` back to a Nanode 1GB so that we can better see the effects of the architectures described in the following chapters.

If the application were not so database-heavy, we should already see some scaling benefits with this architecture. However, even with a database bottleneck, Chapter 6 will look at improvements to the architecture which will give dramatically better scaling abilities. The goal here is simply to see how the architecture works and how to set it up on Linode.

It is also good to see that just because you can add nodes to a system doesn't make it automatically scalable.

CHAPTER 6

Improving Scalability with Caching

This chapter will cover several adjustments to the basic two-tier architectures described in the previous chapter.

6.1 Understanding Caching Architectures

In our current application, the database is the main bottleneck. This means that adding more web servers will not substantially increase the load that the cloud can handle. When you discover a bottleneck, it is good to take some time and think about ways that the bottleneck can be avoided.

In our case, our data doesn't really change all that often. Even if it did, getting up-to-the-second values from a guestbook isn't all that mission critical. If someone had to wait seconds or even minutes to see a recent guestbook entry, it wouldn't be the end of the world.

When you have content that is accessed often, but can afford to be just a little stale (or a lot stale), you can implement *caching* to speed things up. Caching simply means to have a temporary store of results that we can access quickly. Databases are slow because databases are primarily concerned about *data integrity*. You want to know that your data is safe, stored on disk, won't go away, and can be accessed with arbitrary queries. A cache, on the other hand, is ephemeral—they are usually just stored in

J. Bartlett, *Building Scalable PHP Web Applications Using the Cloud*,
https://doi.org/10.1007/978-1-4842-5212-3_6

memory. Their goal is for fast data retrieval at all costs. Many caches, for instance, will just start throwing away data if they fill up their memory. That's fine, because, if something doesn't exist in the cache, we can go to the database and refetch the data.

In short, databases are about permanence and reliability, and caches are about getting what I want as fast as humanly possible. Caches are usually implemented as simple key/value pairs, usually with an expiration date attached. That is, each piece of data that the cache stores has a designated "key." For instance, since there is only one listing of all the guestbook entries, we can have a single cache "key" for this. In the application, this will be called `entrylist`. However, each individual guestbook entry might wind up getting its own cache key (so the cache can know that it is unique), which would include the database ID of the entry. Cache keys are usually just regular strings, and, with most caching systems, the content can be anything at all. Caches do not do any special processing, they just store whatever you ask it to in a given key, so you must be careful not to use the same key for two different things!

The way a cache is used is that the program will first figure out what key to use for the data. The goal is to be able to easily infer the cache key from the URL parameters. The program starts by checking to see if the cache has a value for the cache key. If the cache contains the value, then that value is used. If it doesn't, the program then gets the value the "normal" way (i.e., usually going to the database for the value, and perhaps doing some additional calculations or manipulations) and then saves it to the cache with an expiration date. Then, whether the value came from the cache or the "normal" way, the value is used in the application. Once the expiration date passes (or the cache gets too full of values), then the cached value simply disappears for the next query, forcing the program to get a fresh value the "normal" way.

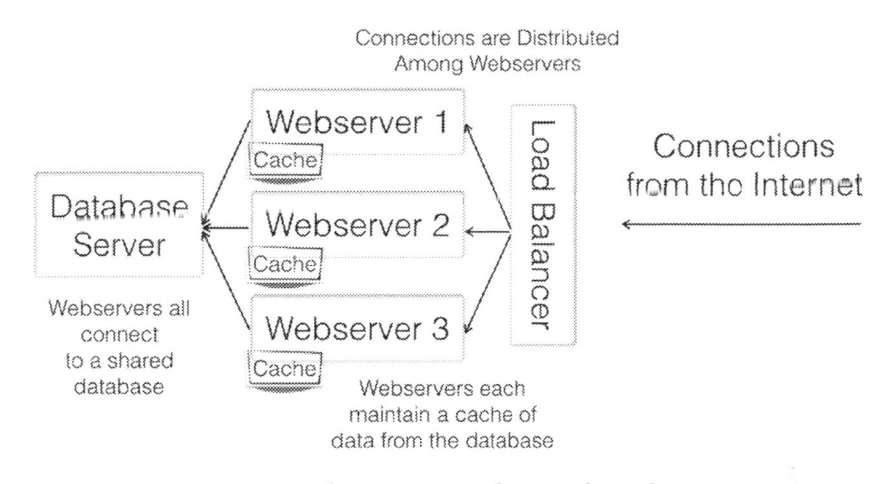

Figure 6-1. *Two-Tier Architecture with Local Caching*

As you can see, this methodology of cache vs. normal accessing makes sure that while the caching mechanism is preferred, the non-cache mechanism is available as well, and this mechanism will fill the cache.

As noted, the database is usually considered the slow part of any application, because it has to be very careful to handle your data correctly and keep it forever. Caches, on the other hand, are considered ephemeral. If we just want to clear the cache all of a sudden, it would not affect the operation of the page, because it would just go back to the database and get the value again.

If a site is heavily loaded and continually asking for the same content, even caching for a few seconds can greatly increase speed and decrease resource usage. We previously load tested this cluster as serving up approximately 250 requests per second on the listing page. If we cached the results of that page for 10 seconds, then that would be 2,500 fewer database requests over that time period!

Figure 6-1 shows how we can modify our standard two-tier architecture to add a caching layer. In this architecture, each individual web server maintains its own cache of results. This can lead to small inconsistencies between web servers since they may have refreshed their results at different times. However, if the expiration is not set too far in the future, these issues are minimal. Additionally, if there are issues, remember in Section 5.6 we discussed load balancer "Stickiness." This ties a specific user to a specific server, which means that user will always be on the same server, which means they will always be using the same server cache. Additionally, this makes the caching more efficient, as each server only has to cache data for a subset of users.

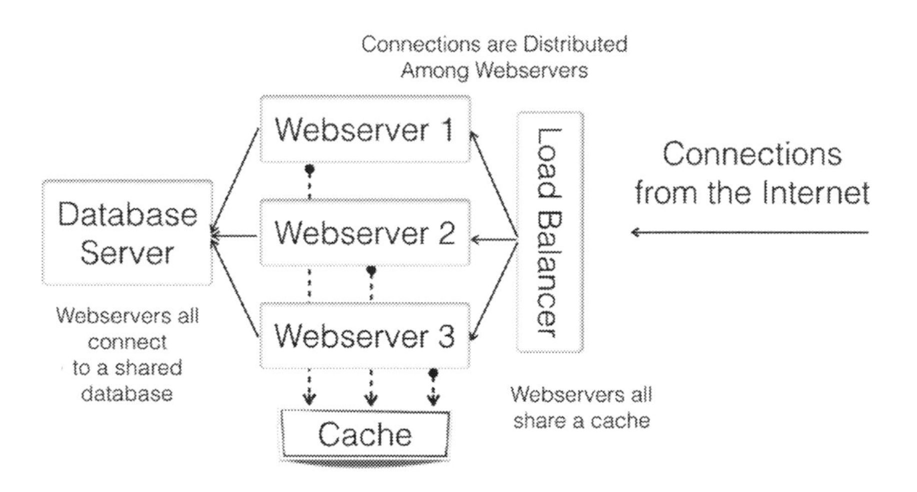

Figure 6-2. *Two-Tier Architecture with Global Caching*

What to cache and how long to cache it is very application dependent. For many architectures, there are pieces which can be cached for minutes or even days, and other pieces that can only be cached for seconds or can't be cached at all. Caching can often lead to unexpected results, so it is good when you are developing to include a feature that can be used to turn off caching altogether in order to see if your cache is the problem.

If keeping the cache consistent among all of your servers is important, another architecture to consider is using a global cache, as in Figure 6-2. In this architecture, rather than each web server maintaining its own cache, there is a shared cache that they all access. This adds a little more network latency since all caching calls have to run over the network, but it adds more consistency to your results.

Depending on how it is implemented, this sort of external cache can also become a bottleneck. However, there are many caching servers that can span across multiple servers and balance requests among the servers. Nonetheless, this also adds a layer of management complexity to the puzzle.

In nearly all cases, I find that having a single, external caching service is difficult to set up and maintain, with little if any benefit. For most situations, putting the cache on each web server gives you the most scalable efficiency, even if it comes at a cost of a little consistency, which, as mentioned, can be alleviated by using load balancer "stickiness."

6.2 Implementing Caching in the Application

The caching architecture we are going to implement here is the one shown in Figure 6-1, both because it is easier to implement and because it is easier to manage in the long run. What we are going to do is perform the configuration change on the `template_node` machine, and then simply turn off our existing `webnodeX` boxes and replace them with new boxes replicated from `template_node`. This is a lot easier (and less error-prone) than going through every server and making the change.

The first thing we need to do is to install the caching service. We will use `memcached` for our caching service because it is straightforward to run, access, and manage. To install `memcached` and turn it on, just enter the following commands as root:

```
yum install -y memcached
systemctl enable memcached
systemctl start memcached
```

You will also need the PHP extensions for memcached. These have to be compiled, so we will need to install a few more extensions (again, as root).

```
function getCache() {
  $conn = new Memcached();
  $conn->addServer("localhost", 11211);
  return $conn;
}
```

Figure 6-3. *Memcache Connection Function*

```
yum install -y libmemcached
yum install -y php74-php-pecl-memcached
```

Don't forget the "d" at the end of memcached, because there is another extension named just memcache which does not do what we want. Note that this also enables the extension in PHP. It does this for us, but if you needed to tweak the PHP side of the configuration, the file is located in the directory /etc/opt/remi/php74/php.d.

Now we need to restart our PHP-FPM process to use the new PHP extensions:

```
systemctl restart php74-php-fpm
```

Next, we need to modify our application to create a connection to our local memcached service. Therefore, add the code from Figure 6-3 into common.php. The number 11211 referred to in the code is the port that memcached is listening on by default.

The way that your code should use the cache is as follows:

1. Create a cache key to uniquely identify the information in the cache.

2. Check to see if the information already exists in the cache. If so, use it.

3. If the information doesn't already exist in the cache, find the information the slow way (i.e., perform the database query).

```
<?php
  include("common.php");
  $cache = getCache();
  $key = "entrylist";
  $val = $cache->get($key);
  if($val) {
    $result = $val;
  } else {
    $dbh = getReadOnlyConnection();
    $stmt = $dbh->prepare(
      "select * from gb_entries" .
      " order by id"
    );
    $stmt->execute();
    $result = $stmt->fetchAll();

    $expiration_seconds = 10;
    $cache->set($key, $result,
       $expiration_seconds);
  }
  getHeader();
?>
```

Figure 6-4. *Rewriting* `list.php` *to Use the Cache*

4. Take the information and store it in the cache using that key with a future expiration time (in this case, we will set the expiration to be 10 seconds from when it was set).

Next, we will update the `list.php` function to use the cache. Figure 6-4 shows how to rewrite `list.php` to use the cache. This is just the top portion of the script which stores the result of the query in `$result`. The actual HTML output code remains the same.

6.3 Reimaging the Cluster

Now we need to deploy this to our cluster. To do this, we will need to first make another snapshot backup of `template_node` and then perform the following steps for every `webnodeX` server in our cluster:

1. Go to the node balancer and remove the `webnodeX` server from the configuration.
2. Power off and delete the `webnodeX` server (deleting can be done from the "Settings" tab of the node's screen).
3. Create a new `webnodeX` server in its place (with the same name) based on our `template_node` backup (be sure to add a private IP). Power it on if necessary.
4. Add the new `webnodeX` server to the node balancer configuration.

For a production system, this is not always the best way to do a code or a system update. We will cover additional ways of doing this in Chapter 12. This method simply removes all of your `webnodeX` servers and replaces them with new ones. It is a pretty clean mechanism, though you would want to perform it slowly to be sure your site stays up while you are reimaging the servers.

6.4 Testing Our Caching Architecture

Once you have your new cloud cluster up and running, it is time to test out the new architecture and see if we made any performance gains.

For this setup, I ran ApacheBench on individual servers and on the balanced cluster. The individual servers, since they no longer relied on the database for a bottleneck, were able to serve up over 800 requests per second!

Also, because there was no bottleneck on the database, the performance was able to scale almost linearly. Linear scaling means that each box you add gives you the same performance boost. In this case, with a single server, our performance was 800 requests per second, two servers yielded around 1,500 requests per second, and at three servers, we were able to consistently serve up over 2,300 requests per second!

Note that if you are not seeing similar performance gains, check your balancer configuration to be sure that you didn't turn on session stickiness at any point, as this will restrict your ApacheBench sessions to a single server. Also, on the "Settings" tab, be sure that the "Client Connection Throttle" isn't turned on, either.

Figure 6-5 shows the output of ApacheBench on the full cluster.

So, what we have learned is that not only did caching improve the speed of the application, it improved the *scalability* of the application. Because we only have to hit the database when the cache expires, the speed of the database is now relatively unimportant. In fact, even if we sped up the database again (like we did at the end of Chapter 5), it would have relatively little impact on our total efficiency, simply because it is rarely used.

On the downside, if you actually use the app, you will find that after you post a guestbook entry, it won't immediately display on the site. In fact, if you reload the page quickly, you might find it appearing and disappearing, depending on when the server you are on refreshes its cache.

This can be mitigated in a variety of ways. First, you can lower the amount of time that the data is being cached for. It makes very little difference in this benchmark whether you cache for 1 second or for 10.

So, just changing the cache expiration line to the following (i.e., setting the expiration time to 1 second instead of 10) will make the app very quickly refresh without significantly impacting the performance on these massive loads we are testing against:

```
$expiration_seconds = 1;
```

```
Server Software:        Apache/2.4.6
Server Hostname:        BALANCER.IP.ADDRESS.HERE
Server Port:            80

Document Path:          /list.php
Document Length:        383 bytes

Concurrency Level:      50
Time taken for tests:   0.425 seconds
Complete requests:      1000
Failed requests:        0
Write errors:           0
Total transferred:      560000 bytes
HTML transferred:       383000 bytes
Requests per second:    2353.76 [#/sec] (mean)
Time per request:       21.243 [ms] (mean)
Time per request:       0.425 [ms] (mean,
   across all concurrent requests)
Transfer rate:          1287.21 [Kbytes/sec]
   received

Connection Times (ms)
              min  mean[+/-sd] median   max
Connect:        0    0   0.4      0       4
Processing:     2   19  13.4     14     108
Waiting:        2   19  13.4     14     108
Total:          2   20  13.4     14     109
```

Figure 6-5. *ApacheBench Output for Cached Configuration*

Even more, if you keep sessions tied to a particular server, you can actually tell the server to clear individual keys or even the whole cache on certain events. Therefore, at the end of `create.php`, we could add the following lines to clear out the list cache on that server:

```
$cache = getCache();
$cache->delete("entrylist");
```

Or, if an app was complex enough that there were a whole lot of keys that had to be deleted, the code could just clear out the whole cache altogether with `$cache->flush();`.

In any case, as you can see, caching architectures can make your app slightly more complex and difficult to manage, but they are usually worth it from the often dramatic performance and scalability boost.

CACHE DEBUGGING TIPS

Caches, while hugely beneficial, bring in their own set of problems. In order to better debug web pages, it is best if you always have a set of parameters that you can pass to the application to get it to turn off caching. For instance, in many of my own applications, passing in a `no_cache=1` in the URL will turn off caching. This is usually the first place I go to when problems are reported.

Here are some signs that you might be running into caching problems, and what to do about them:

- **Problem**: Your application is putting out old content, even though newer content is in the database.

  Diagnosis: The cache is holding on to stale content.

  Solutions: Expire your content sooner, or provide additional cache key information to let the cache know when you need new data.

- **Problem**: Your application is spitting out inappropriate data given the parameters.

  Diagnosis: This often happens when your cache key is not specific enough. For instance, if content was coming out in the wrong language, that might mean that you need to attach the current language as part of the cache key.

  Solutions: Add more parameters to your cache key to make sure you truly identify each piece of unique content with a unique key, or you may decide that this content it too specific to be cached.

- **Problem**: The same page is spitting out different content on each reload.

  Diagnosis: The cache has different content on each server depending on when it was accessed.

  Solutions: There are several ways to fix this. You can (a) decrease the amount of time until expiration, (b) increase the "stickiness" on the load balancer to be sure that the same person is always hitting the same server (and thus the same cache), (c) utilize a global (or synchronized) cache instead of a local cache, and (d) add additional cache key parameters to better coordinate what data a user gets from the cache.

- **Problem**: The cache is not speeding up your application as much as you thought.

  Diagnosis: Either you are never hitting your cache or you are not caching the right things.

  Solutions: As there are many ways this can go wrong, there are many ways to fix it. Often this occurs when each user is accessing a different set of data, and therefore nothing winds up being pulled from the cache. This can be fixed by increasing the cache size and/or intelligently preloading the cache with data that is likely to be accessed. You may need to increase the expiration time of your data as well. However, it could be that you are having to do a lot of postprocessing of the data, and this is taking longer than the actual query.

CHAPTER 7

Database Replication

Some things simply cannot be cached. Ad hoc reports, up-to-the-second changes, and sites where access patterns are spread across a large number of unrelated pages are all difficult to optimize using caching. For workloads like this, you *can* deploy a bigger database server, but eventually even those run into limits.

Therefore, many application architectures include database replication, where there is more than one database server serving out requests.

7.1 Types of Database Replication

There are many types of database replication depending on your needs. The basic types of replication include

> **Failover Replication**: In this configuration, the replicated servers do not help with the load, but they do make sure that if the primary database server goes down, there is a database with up-to-date information ready to take over.
>
> **Master/Replica Replication**: In this configuration, the master database is the only one with read/write access. The replica servers receive data as it is recorded on the master (or shortly thereafter),

J. Bartlett, *Building Scalable PHP Web Applications Using the Cloud*,
https://doi.org/10.1007/978-1-4842-5212-3_7

but are read-only copies of the master database. All updates go to the master database, but queries can go to either the master or any replica server. This is also known as "master/slave" replication, where the replica database is considered the "slave database."

Multimaster Replication: In this configuration, all databases are considered to be equally "master" databases, and writes can be performed on any of them. Writes to any given database are then synchronized with the rest of the cluster.

This chapter will focus on master/replica replication because it is the easiest to implement, has the fewest practical problems, and gives the most results for the least effort. Multimaster replication is rarely used because it is difficult to set up, maintain, and keep efficient, and very few databases support it. Even when supported, multimaster replication often introduces new, difficult to solve problems, such as data conflicts (i.e., when conflicted data is committed on two different servers). Therefore, to maintain simplicity, this book will focus on master/replica configurations.

Figure 7-1 shows a conceptual view of a typical master/replica architecture.

7.2 Replicating the PostgreSQL Database

PostgreSQL's replication system has advanced quite a bit over the years both in features and in ease of use. While it is not *difficult* to use, it does take some explanation to understand.

The built-in PostgreSQL replication system uses a technique known as log streaming to replicate. PostgreSQL, in order to guarantee data consistency, creates what is known as a write-ahead log, or WAL. Basically, PostgreSQL writes to the WAL the changes it is *about* to do and then actually does the changes. This means that if the database server powers off during an update, it has a record of what it was doing and can simply finish the operation when it comes back on.

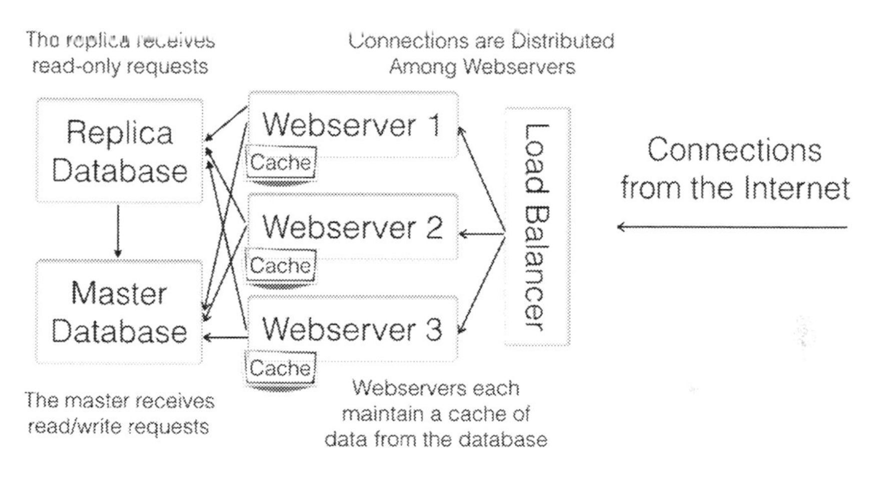

***Figure 7-1.** Diagram of a Master/Replica Database Architecture*

Interestingly, this is precisely the information that a replication server would also need to know. Therefore, to implement database replication, PostgreSQL simply ships the WAL files to the replication servers, which likewise implement the changes. This type of replication is known as WAL streaming.

To accomplish this in our cluster, we need to configure our main database in order to receive replication connections. Log in to `dbmaster` as the root user, edit the file `/var/lib/pgsql/data/postgresql.conf`, and set the following settings:

```
wal_level = hot_standby
wal_keep_segments = 32
max_wal_senders = 4
hot_standby = on
```

If you are using a later version (9.4 or greater) of PostgreSQL, you will also need to set:

```
max_replication_slots = 4
```

However, this setting will break the version of PostgreSQL that ships with CentOS 7.2 that we are working with in this book.

These configuration changes accomplish several things:

- `wal_level` adjusts PostgreSQL's "write-ahead log" (i.e., WAL) so that it retains enough detail to send everything that a replica server would need.
- `wal_keep_segments` keeps enough data from the WAL hanging around after it is used so that the replica server can still get to the data if it falls behind. We have set this to 32, which is a pretty conservative setting. This allows for a new replica to have a lot of time to get fully synchronized and prevents minor network glitches and slowdowns from desynchronizing the servers.
- `max_wal_senders` should be set to at least the number of replica servers plus two (and, if you decide to install a newer PostgreSQL, you may need to set `max_replication_slots` to the same value).
- `hot_standby = on` allows the replica server to respond to queries.

Next, we need to add a replicator user to the PostgreSQL database. Type in `psql -U postgres` to access the database, then type the following to create the user `replicator` for replication (all on one line):

```
CREATE ROLE replicator WITH REPLICATION
PASSWORD 'mypassword' LOGIN;
```

Then type `\q` to exit.

Next, we need to give remote servers permission to open replication connections to this server. Add the following line to /var/lib/pgsql/data/pg_hba.conf:

```
host replication replicator all md5
```

Now we need to restart the database so that the new settings take effect:

```
systemctl restart postgresql
```

Our server is now completely ready to start taking replication requests. Now we can set up the replica server.

In order to set up a PostgreSQL replica server, the first thing we need to do is to create a new Linode node for it to run on by replicating our template node using the standard procedure. For this exercise, name the new node dbreplica. You will also need to add a private IP address to this machine and write it down (we will refer to it as DB.REPLICA.PRIVATE.IP for the rest of this book).

Now, boot up dbreplica and log in as root.

If you have followed the instructions, this machine should not have PostgreSQL running. If it is running, however, you can turn it off with systemctl stop postgresql. Once PostgreSQL is off, we need to clear out the existing PostgreSQL installation. Do that with the following command:

```
rm -rf /var/lib/pgsql/data/*
```

Now we need to request an initial binary backup from the master system as a starting point for replication. This is accomplished with the pg_basebackup command. To generate this initial backup starting point, first switch to the postgres user like this:

```
su - postgres
```

Next, issue the following command (all on one line):

```
pg_basebackup -x -U replicator -h DB.MASTER.PRIVATE.IP
 -D /var/lib/pgsql/data
```

It will ask you for the password, and then it will copy the entire PostgreSQL instance from the master database, including the configuration files. Next, you will need to tweak the configuration files after this step is complete. The `postgresql.conf` file that we set up in Chapter 5 has a command `listen_addresses` which includes the server's private IP address. Unfortunately, since this was copied from the master database, it currently has the master database's private IP address. To fix that, just open up `/var/lib/pgsql/data/postgresql.conf` and change the `listen_addresses` configuration to read:

```
listen_addresses = 'localhost,DB.REPLICA.PRIVATE.IP'
```

After this concludes successfully, you need to tell PostgreSQL that this server is to be used as a hot standby.

This is done by telling the server to go into "continuous recovery mode" on startup. To do this, we create the file `/var/lib/pgsql/data/recovery.conf` with the following contents (the last two lines should both be typed on the same line):

```
standby_mode = 'on'
primary_conninfo = 'host=DB.MASTER.PRIVATE.IP port=5432
 user=replicator password=mypassword'
```

Once this is all in place, you need to exit back to the root user like this:

```
exit
```

Now that you are the root user again, we need to turn the PostgreSQL database back on and make sure it will come on automatically if it reboots:

```
systemctl start postgresql
systemctl enable postgresql
```

We also need to make sure there is a hole in the firewall to receive database connections:

```
firewall-cmd --add-service postgresql
firewall-cmd --add-service postgresql --permanent
```

At this point, your system is now up and running as a replica server!

To verify this, run the following command, which will list all of your running PostgreSQL processes:

```
ps afxw|grep postgres
```

One of the processes in the list should include the word `recovering` or startup in the output. This means that the database is up, active, and feeding off of the WAL logs of the master.

You can also check the log files, which will be in the directory `/var/lib/pgsql/data/pg_log`. The end of the log file should say something like "database system is ready to accept read only connections" and "streaming replication successfully connected to primary."

A FEW NOTES ON POSTGRESQL REPLICATION

PostgreSQL implementations are called "instances" and can contain any number of databases. If you are following the instructions in this book, your PostgreSQL instance only contains one database (actually three, since PostgreSQL always comes with a pair of template databases installed, `template0` and `template1`). It is easy enough to create a new database using the `createdb` command-line program or issuing a `create database` instruction while running `psql`.

In any case, keep in mind that since the PostgreSQL WAL file is a system-level feature (i.e., the WAL files are shared by the entire database instance), PostgreSQL streaming replication replicates the entire PostgreSQL instance, not just a single database.

7.3 Setting Up the Application to Utilize Master/Replica Replication

The application itself is already built to make use of read-only replicas. If you remember, we actually have *two* connection functions, `getReadWriteConnection()` and `getReadOnlyConnection()`. Right now, they are both pointing to the same server. To make use of our new read-only replica, all we have to do is change the connection information in the `getReadOnlyConnection()` function, and it will shift all read-only connections onto the replica server.

All we have to do is change the `host` parameter in the `getReadOnlyConnection()` function to point to `DB.REPLICA.PRIVATE.IP`.

Once this is done, we just need to redeploy our code. As usual, we can do this either by deploying to `template_node` and then re-creating the `webnode` servers from it or by deploying it to each server individually.

7.4 Adding More PostgreSQL Replica Servers

If separating out a single replica server does not give you the performance increase that you need, you can actually have as many replica servers as you need. You can do this by either replicating the template node and repeating the process in Section 7.2 or by directly replicating the replica server.

Replicating the replica server is not quite as automatic as replicating the web nodes, but it can save a few steps. In order to do this, you need to first enable backups on your current `dbreplica` node and then create new replica instances from the backups of `dbreplica`. After you create each replica instance, you need to `ssh` into the new replica server and set the `listen_addresses` value in `/var/lib/pgsql/data/postgresql.conf` to its own private IP address, because it will be set to the one for `dbreplica` by default. After doing this, you will need to restart PostgreSQL with `systemctl restart postgresql`.

After you create a number of copies of your database, you need to modify your code to make use of your databases. It would be nice if we could create a load balancer for our replica databases and then just point all of our application code to the load balancer. Unfortunately, Linode does not currently have the ability to create *internal* load balancers (i.e., a load balancer that only accepted requests on the private network), so we will have to provide our own load distribution mechanism. We will instead emulate this feature in the application code by selecting a server at random when getting database connections. Because we don't have a load balancer, we will have to hard-code the list of servers in your application code, and adding a new replica server will also require modifying the application code and repushing that code to all of the web servers.

To understand how the application code will be modified, let's say that we now have three replica servers, with the private IP addresses of `DB.REPLICA.PRIVATE.IP`, `DB.REPLICA2.PRIVATE.IP`, and `DB.REPLICA3.PRIVATE.IP`. To get our read-only connection to cycle between these, we will rewrite our `getReadOnlyConnection()` function according to Figure 7-2.

Once this code is in place in our cluster, all of our read-only requests will be load balanced across multiple replica servers like it shows in Figure 7-3. This will mean that our only bottleneck is on read-write database requests. Most applications are dominated by read-only requests anyway, so having a bottleneck on read-write requests is usually unproblematic.

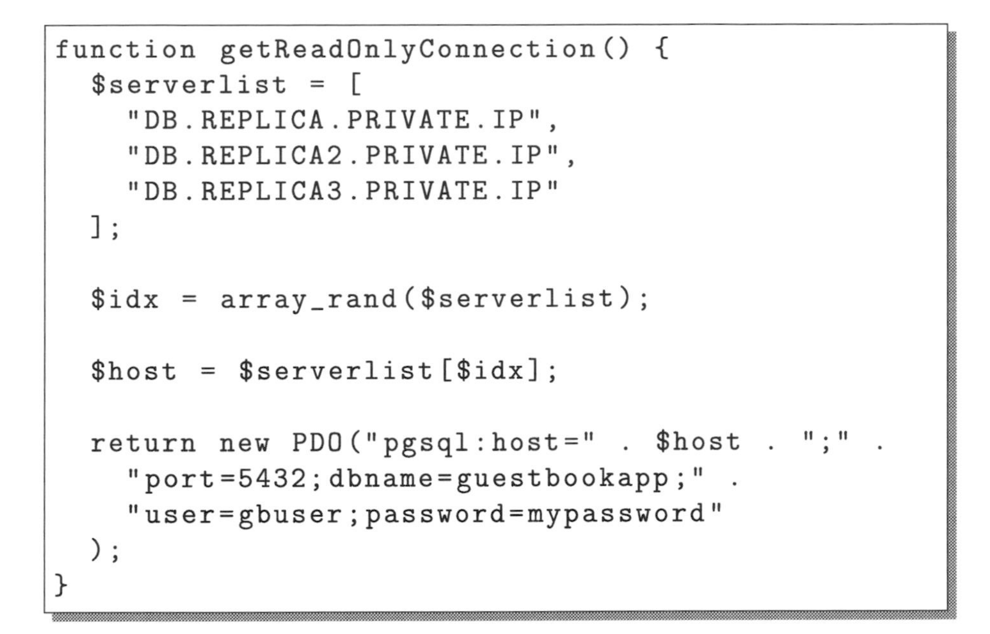

```
function getReadOnlyConnection() {
  $serverlist = [
    "DB.REPLICA.PRIVATE.IP",
    "DB.REPLICA2.PRIVATE.IP",
    "DB.REPLICA3.PRIVATE.IP"
  ];

  $idx = array_rand($serverlist);

  $host = $serverlist[$idx];

  return new PDO("pgsql:host=" . $host . ";" .
    "port=5432;dbname=guestbookapp;" .
    "user=gbuser;password=mypassword"
  );
}
```

***Figure 7-2.** Connecting to a Group of Replica Servers*

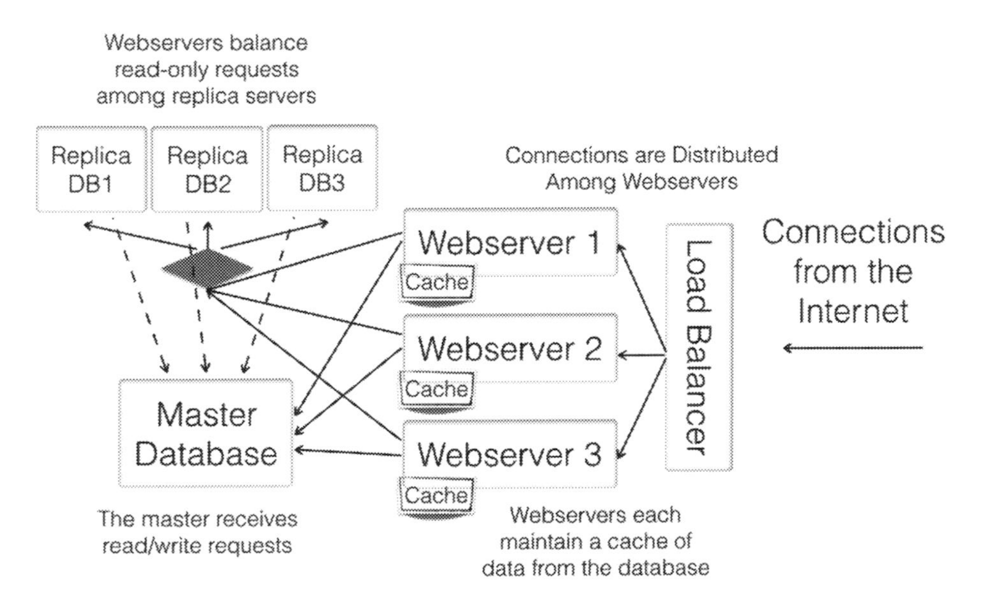

***Figure 7-3.** Diagram of a Multi-Replica Database Configuration*

7.5 Replicating Across Datacenters

For very large applications, sometimes you want better geographic coverage than can be provided in a single datacenter. Additionally, ultracritical applications may need the reliability that comes from having multiple datacenters so that if one datacenter goes down, the application can continue to at least partially function.

The creation of this sort of setup is a bit more involved, so this book will not present a step-by-step method as has been the case for other architectures. However, these are the basic steps you will need to perform to achieve this functionality:

1. Change `dbmaster` so that its `listen_address` is set to *. Since it will be receiving requests from other datacenters, it has to listen on the public IP address as well.

2. Firewall `dbmaster` to prevent access from unwanted third parties.

3. Encrypt your connections to PostgreSQL using SSL (see later how to enable this).

4. Deploy a "primary replica" in the new datacenter, which will be a replica to `dbmaster` but will also be a master for other replicas on the network if you need them. This is still read-only, but it is a replica that streams to other replicas.

5. Deploy a copy of your web application to the new datacenter, which has the public IP address of `dbmaster` for read-write connections and the list of private IP addresses of the local replicas for read-only connections.

At the end of the process, your cluster architecture should look something like Figure 7-4.

As your deployment becomes more complex, your need for automated deployment handling increases. See Chapter 12 for more information on automated deployments.

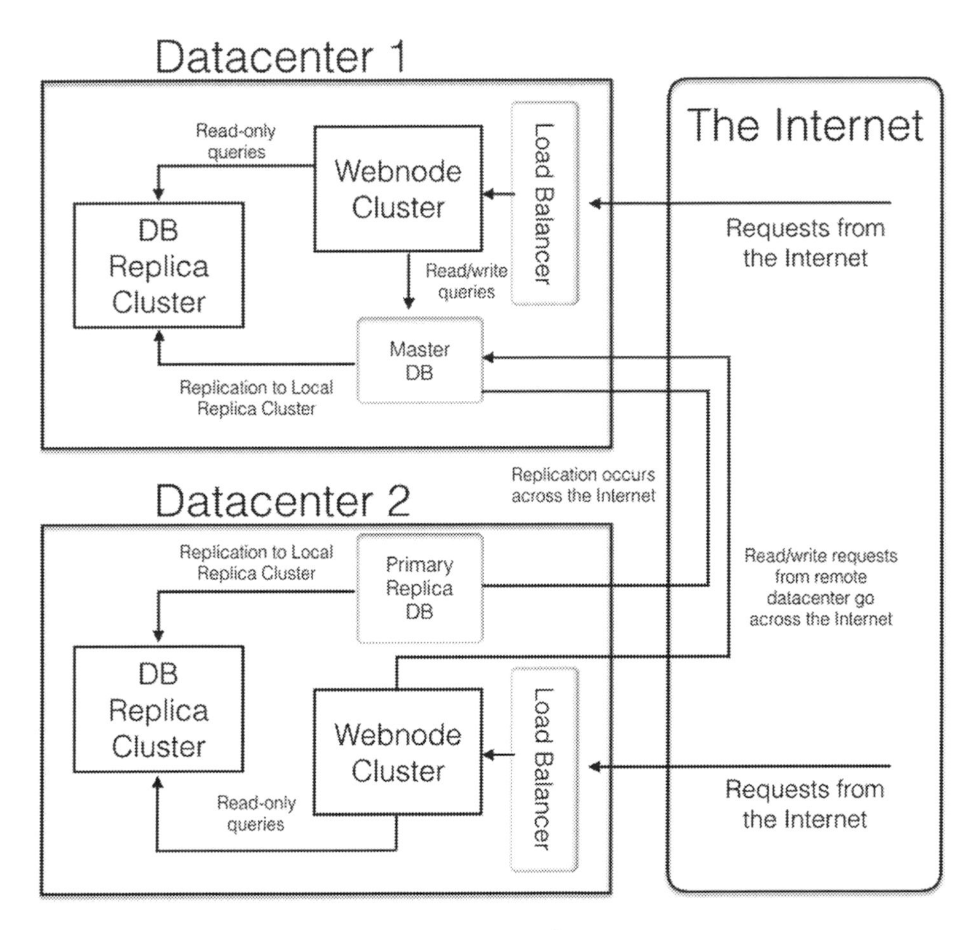

***Figure 7-4.** Diagram of a Multisite Architecture*

If you need read-write access to the database within your local setup, the configuration becomes still harder, as you now have to manage how synchronization occurs between your systems and what happens if the network connection between them goes down. In cases like this, you often separate out local (unshared) and global (shared) data into *separate* database instances. For the unshared data, you will want a master database at each location. For the shared data, you will set it up similar to Figure 7-4.

ENABLING ENCRYPTION ON POSTGRESQL

To enable encryption on PostgreSQL, you need to go to the directory `/var/lib/pgsql/data` as the root user. From there you need to edit `postgresql.conf` and add the following line:

```
ssl = on
```

Next, you need to generate an SSL key and self-signed certificate. From that same directory, issue the following command (all on one line):

```
openssl req -nodes -new -x509 -keyout server.key
  -out server.crt
```

This will require you to fill out a short form which will be encoded into your certificate. Then, you need to set the permissions on the generated files:

```
chown postgres:postgres server.key server.crt
chmod 600 server.key
```

If you restart the database at this point, the server will *allow* SSL connections but not *require* them. To require SSL connections, you can change any or all of the `host` lines in `pg_hba.conf` to be `hostssl` instead.

Once all of your changes are made, you can restart the database with:

```
systemctl restart postgresql
```

7.6 Sharding Your Data

Another option for database scalability is data *sharding*. Database sharding is relatively simple in concept—it just means that you partition your data in such a way that not all of the data relies on the same database system.

For instance, you could shard your data so that all customers whose names start with "A" and "B" are in one database, all customers whose names start with "C" and "D" are in another, and so forth. You can partition the data on whatever method you prefer, as long as the system can easily determine which database it needs to query. The primary point is that rather than everything being managed by the same database system, the data itself is split out to different database systems.

In such systems, usually either the application takes care of sharding or there is a middle layer that handles handing off connections to the appropriate databases. Some newer tools have been developed, such as `pg_shard`, which aims to operate seamlessly as a database plugin.

In any case, sharding brings with it a whole set of data management and data integrity issues. One of the points of databases was to ensure consistency of data. Sharding, essentially, removes many of the protections that were offered by databases in order to achieve scale. Therefore, it is important to be sure you know why you want to shard and how you want to shard in order to minimize risk.

For instance, if you sharded customers to different databases, you should also shard related records as well, so the customer is serviced by a single database, and so that their records are managed together as well. Imagine if you had to restore records from a backup, but some related records were on a different database system!

Sharding is a lot of work, takes a lot of planning, and is heavily dependent on the specifics of your usage and workload, so it is hard to speak of it in generalities. Nonetheless, if you are looking for more ways to scale your database, sharding is definitely an option. Sharding also can work in conjunction with other methods, such as master-replica replication or multimaster replication, but again, it requires a lot of extra love, care, and management to make it work well.

Sharding is easier if your login groups don't share any data. For instance, let's say that you built an e-mail marketing system. Different customers rarely share any data on such a system. Therefore, it is no problem to keep their records completely separate.

You could potentially run several entirely independent clouds, each one with a different set of customers. Logins from customers A, B, C, and D would get shuffled into Cloud 1, logins from customers E, F, G, and H would get shuffled into Cloud 2, and so forth. If these clouds are all managed completely separately, such a setup would also allow for some level of disaster mitigation, as you would be unlikely to have a complete outage of all datacenters at the same time.

CHAPTER 8

Using a Content Delivery Network

So far, the focus of our scalability efforts for our cloud application has been the serving of dynamic content (i.e., content pulled from database queries). However, for many sites, dynamic content is actually the smallest part of their system. In fact, on any web site, most of the requests are not even for your dynamic content, they are for your static content—your images, stylesheets, and JavaScript. Therefore, when looking at ways to scale your application in the cloud, it is important not to forget to also scale your static assets.

You can scale your static assets by creating more front-end web servers, since they are usually the ones to serve static assets anyway. However, this can get more expensive (you have to keep more servers online) and harder to manage (more nodes equals more management).

Scaling static assets is actually much easier than scaling dynamic assets, since you don't have to worry as much about what happens when they change. Therefore, there are services that are built specifically for scaling static assets. A service which scales your static asset delivery is known as a "Content Delivery Network," also known as a CDN.

J. Bartlett, *Building Scalable PHP Web Applications Using the Cloud*,
https://doi.org/10.1007/978-1-4842-5212-3_8

8.1 How Does a CDN Work?

The way most CDNs work is that they have a *globally* distributed network of content servers. Let's say that you have a 10MB image that you want to be served by a CDN. Normally, if that user asked your server for the image, then your server's processing time and bandwidth would be stuck sending that 10MB image. If the user lived on the other side of the ocean, that is an additional problem, as your server will waste a lot of resources managing a slow and noisy connection.

What a CDN allows you to do is to redirect your users to the URL of the image at the CDN. The CDN, the first time it sees the URL, usually doesn't know anything about the image, but it has rules that tell it how to find the original image on your service. From that point on, after the CDN grabs the image the first time, every other time someone asks for that image, the CDN will manage delivering the image to the users without going through your servers at all.

Additionally, most CDNs have servers located in many different physical localities. These locations are known as "Points of Presence" (PoPs), and the servers there are often referred to as "edge servers" (because of this, CDNs are often referred to as "edge caches"). Using edge servers at a variety of PoPs allows a CDN to not only provide scalability through a large number of servers, edge servers allows a CDN to provide servers that are *in close physical proximity* to the user. This greatly speeds up your user's experience if they can get a lot of the data from nearby servers, rather than having to cross oceans to retrieve the data.

CDNs provide essentially infinite scale for serving static assets—for any big-name CDN, you don't have to worry about overrunning their network. As long as the asset isn't changing, the CDN will be perfectly able to handle any number of requests for that asset.

Usually, the primary cost for a CDN is the bandwidth. However, many CDNs have lower bandwidth costs than cloud servers do. Now, with Linode, you would have to have an extremely active site to outrun your included bandwidth. Nonetheless, if you did, you would pay slightly less for the bandwidth if it was provided by a CDN. In any case, even when you have plenty of included bandwidth with your server, the benefits of improved scalability with CDNs are usually worth their small price, as you don't need to be continually running servers to deal with traffic that may or may not show up. You only have to pay for the bandwidth that you actually use.

8.2 Setting Up a Simple CDN

Thankfully, most CDNs are extremely easy to set up. In this book, we will use Amazon's CloudFront as the CDN, though many other options exist (CloudFlare, StackPath, CDN77, and Fastly, just to name a few). The nice thing about CDNs is that, since the service they provide is fairly transparent, it is easy to mix and match service providers for CDNs.

The way that CloudFront works is very simple:

1. You host all of the static content on your main site. This is the "official" repository of your content.
2. You create a host on CloudFront to serve your content from (usually this will have a name such as `xyzabc.cloudfront.net`).
3. You tell the CloudFront server the URL of your main site.

4. Every time you link to static content from your site, you link to it using the CloudFront URL, rather than linking to the content on your own server. For instance, if your image was at `http://mysite.example.com/mydirectory/myimage.png`, when you linked to it, you would use the URL `http://xyzabc.cloudfront.net/mydirectory/myimage.png`. CloudFront will know from your configuration how to find `myimage.png` and will cache it and serve it to your users.

The first time that CloudFront receives the request for the image, it will go to your site to grab it. Then, going forward, any future requests will be served directly by CloudFront using a server near to the user requesting the image.

Let's look at how to actually do this using Amazon AWS and CloudFront. The first thing you will need to do is sign up for AWS at `http://aws.amazon.com`. I am going to assume that you can do the signup process without me.

AWS has an overwhelming number of available services, such that the dashboard, rather than listing them all, has you search for one. Type "CloudFront" into the search bar, and it will then allow you to proceed to the CloudFront Dashboard.

Since this is your first time to use CloudFront, it will simply give you a button which says, "Create Distribution." In CloudFront terms, a *distribution* is simply a CDN replicator service.

After you click "Create Distribution," CloudFront will ask you for your delivery method and give you a choice of "Web" or "RTMP." RTMP is the protocol used for streaming large amounts of video content. However, since we are just distributing basic assets like images and stylesheets, we will choose to get started with basic Web delivery.

Amazon then asks for details about your distribution, as shown in Figure 8-1. There are actually a host of additional options below these, but the only required field is the "Origin Domain Name," which is the DNS hostname of the site you want CloudFront to get its assets from. This *cannot* be an IP address, but must be a DNS hostname of some sort. If you have not set up a DNS hostname for your application, you can just use one of the automatically generated ones from Linode. If you go to the Node Balancer screen in Linode and click your balancer, it will give you an (very long, possibly split across two lines) internally generated hostname for your balancer (something like `nb-BALANCER-PUBLIC-IP-ADDRESS.dallas.nodebalancer.linode.com`). After filling out the "Origin Domain Name," scroll way down to the bottom of the page.

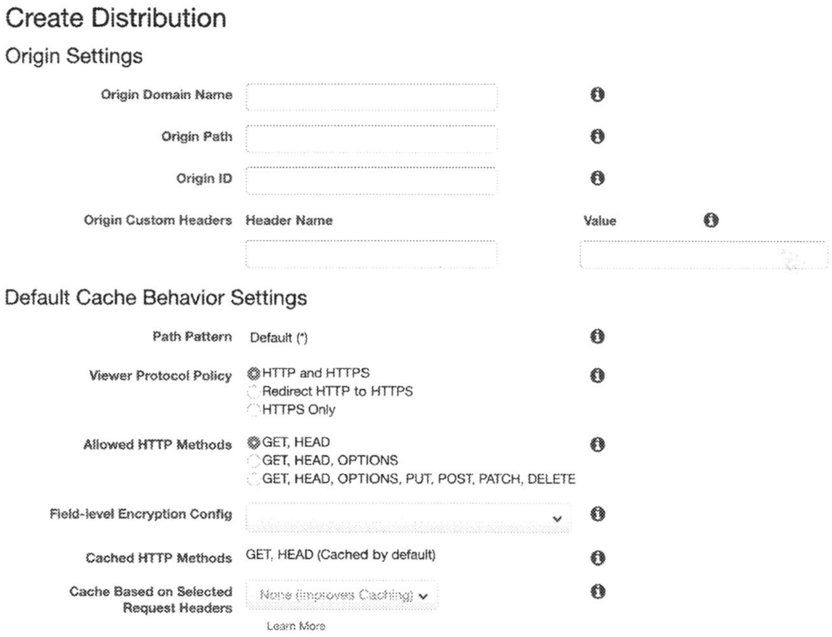

Create Distribution
Origin Settings
Origin Domain Name
Origin Path
Origin ID
Origin Custom Headers Header Name Value
Default Cache Behavior Settings
Path Pattern Default (*)
Viewer Protocol Policy HTTP and HTTPS / Redirect HTTP to HTTPS / HTTPS Only
Allowed HTTP Methods GET, HEAD / GET, HEAD, OPTIONS / GET, HEAD, OPTIONS, PUT, POST, PATCH, DELETE
Field-level Encryption Config
Cached HTTP Methods GET, HEAD (Cached by default)
Cache Based on Selected Request Headers None (Improves Caching)
Learn More

Figure 8-1. *Creating a CloudFront Distribution*

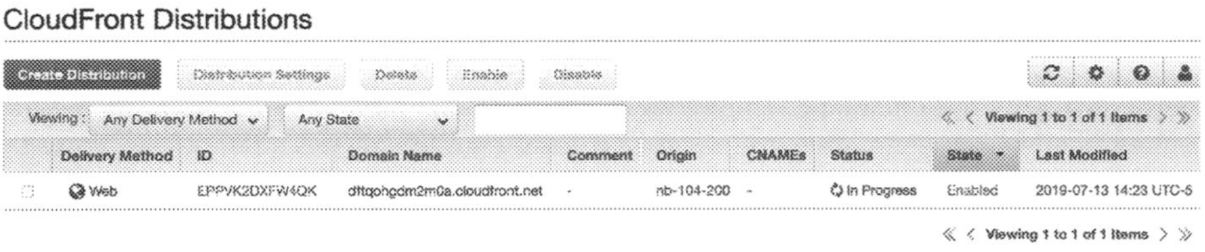

CloudFront Distributions

	Delivery Method	ID	Domain Name	Comment	Origin	CNAMEs	Status	State	Last Modified
	Web	EPPVK2DXFW4QK	dttqohgdm2m0a.cloudfront.net	-	nb-104-200	-	In Progress	Enabled	2019-07-13 14:23 UTC-5

Figure 8-2. *List of CloudFront Distributions*

At the bottom of the page, there is a "Create Distribution" button. After you click this button, you will land in a list of distributions showing your new distribution, similar to Figure 8-2. The most important part of this screen is the "Domain Name," which shows you how to access your newly created distribution (you may need to adjust the size of the field in order to see the whole name). It will also give you a status, which takes anywhere from 5 minutes to an hour to move from "In Progress" to "Deployed." Once it is deployed, you have a CDN up and running!

8.3 Using Your CDN

Once the status has changed to "Deployed," this domain name will now fully mirror your original site. However, it will only be a static version of your site. If your site changes, CloudFront *will not* update its assets unless you tell it to.

Let's say that the domain name that CloudFront gave you was `xyzabc.cloudfront.net`. This means that if you go to `http://xyzabc.cloudfront.net/list.php`, it will show you your guestbook list. However, if you then go and modify your guestbook list, those changes will not be reflected on your CDN—it treats everything as a static asset. That's why, for the most part, CDNs serve up only static assets—images, stylesheets, JavaScript, and the like.

Therefore, instead of accessing the whole site through the CDN, let's modify the application to serve up just our stylesheet through the CDN. All we have to do is modify one line of common.php. In the getHeader() function, we simply need to change the <link> tag to read:

```
<link rel="stylesheet" href="http://xyzabc.cloudfront.net/
guestbook.css" />
```

Be sure to replace xyzabc.cloudfront.net with the domain name of your distribution! Once this is deployed on all of the servers, then your stylesheet will now be served from the CDN.

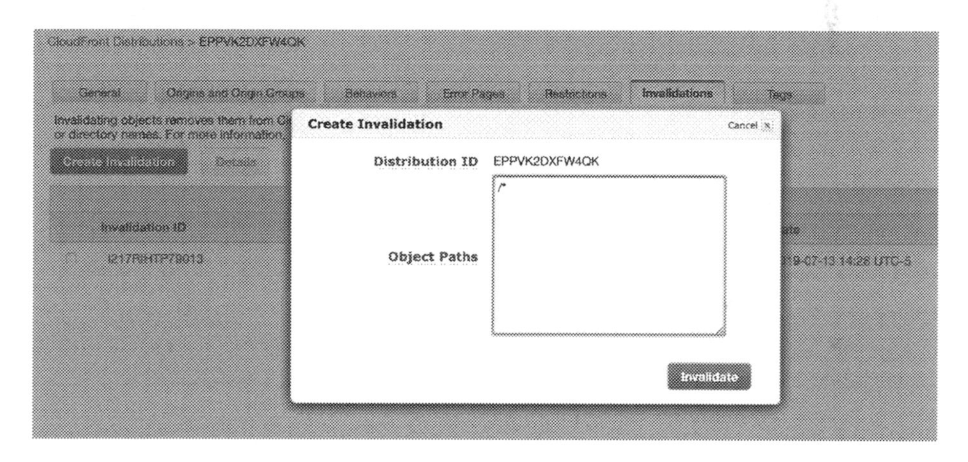

Figure 8-3. *Removing Content from the CDN*

This means that your server will almost never again serve up the stylesheet. Occasionally, the CDN may expire some of its cached content, but that is up to the CDN. The CDN will optimize for itself how much data it stores for how long, and how often it re-requests your original files. For more advanced applications, you can configure Apache to supply either an Expires: HTTP header or a Cache-Control: HTTP header to specify a maximum amount of time that the CDN should hold your data.

However, let's say that you deployed a new version of your application which actually did have an updated stylesheet. This means that the version the CDN is serving up for you is now out of date—it probably still has the old version in its cache. This is not a problem at all, it just means that you need to manually tell the CDN to "invalidate" your content so that it will request it again.

To invalidate content on the CloudFront CDN, start by clicking the ID of your CloudFront distribution. This will bring you to an information page describing all of the options that are set on your distribution. Toward the far right, there is a tab titled "Invalidations." Click this tab, and then click "Create Invalidation." This will bring you to a screen similar to Figure 8-3. In the "Object Paths" field, just clear out anything that is there, and type only /* to invalidate everything. While CloudFront allows you fine-grained access to invalidate and remove specific items from the CDN, I find that, for most situations, simply invalidating the whole thing is easier and cleaner. Click the "Invalidate" button to make it so.

It may take several minutes for the invalidation to spread across all of CloudFront's servers, but in short order all of the servers will be serving your new content. You will know when it is done when the "Status" of the invalidation changes to "Completed."

8.4 Caching Your Whole Site with a CDN

In addition to caching individual pieces of content like stylesheets, JavaScript, and images, modern CDNs can actually allow you to cache your entire web site, giving you instant scalability across the Internet. While this doesn't work for web *applications* such as ours, if you had a basic web site, this could make your web site instantly scalable to an infinite number of users with almost no cost or extra configuration.

Let's look at how to do this with CloudFront. In order for the CDN to serve up your web site directly, you will need to point the DNS of your main site at the CDN's server.

You might think that you could just set the A record of your web site to the CDN. However, CDNs don't typically give you IP addresses for their hosts. The reason for this is that CDNs have *multiple* IP addresses for the CDN (one for each PoP), and they often rely on the DNS lookup to decide which IP address to give to which client (i.e., it will give the client an IP address which is geographically near to them). This means that we can't just specify the IP address of the CDN in our DNS to point the web site to it.

Instead, CDNs usually handle this sort of thing through DNS CNAME records. A CNAME is a "canonical name"—it tells browsers to use a different name for DNS queries and use the results of that name for our own DNS lookups. If you set `www` to be a CNAME to your CloudFront host, then CloudFront can still use its own DNS mechanism to give out the right IP address to each client.

However, this creates another problem. The CDN needs to know what hostname you may be accessing it through. That is, if you have `www.example.com` CNAMEd to `abcxyz.cloudfront.com`, CloudFront needs to know that when one of its machines receive a request for `www.example.com`, it should serve up the cache associated with `abcxyz.cloudfront.com`.

This is done through the distribution settings in AWS. These settings are under the "General" tab of your distribution. Click "Edit" and look for a field called "Alternate Domain Names (CNAMEs)." Here you can put any hostname that you have CNAMEd to this distribution (one per line or comma separated).

This, however, poses a problem, because the CDN needs a place to retrieve your data from, and you just pointed your DNS to the CDN instead of your own servers! To get around this, you simply set up an internal DNS name that the CDN pulls from (you can call this `www2.example.com` or www-internal.example.com), and then set the DNS for www to point to the CDN (note that we are not using "internal" to mean only visible by us, since it is just as much on the public Internet as the main site, but rather to indicate that it is not the destination we are directing end users to).

Doing this, the CDN allows for limitless growth of visitors to your web site.

DEALING WITH BARE DOMAIN NAMES

On a technical note, you cannot create a CNAME for the root-level domain. You can do a CNAME for www.example.com, but not example.com. While your DNS server may allow it, it is against the specification and can cause a variety of strange problems for clients that are not expecting this. Therefore, you will need to make sure you have a mechanism redirecting your bare domain name (i.e., example.com) to your CNAMEd host (i.e., www.example.com).

To get around this, many DNS providers provide a redirect service to automatically redirect all requests for the bare domain name to the www host. If your DNS provider doesn't provide this service, the kind folks at wwwizer.com provide a free service that does this for you. Essentially, if you point the A record for your bare domain to 174.129.25.170, it will be automatically redirected to the same domain with www in front of it.

More information about this service is available at http://wwwizer.com/naked-domain-redirect. Also remember that, as with any overly cheap service, you should take appropriate cautions. Letting them redirect your traffic can simplify things, but letting a third party that you don't have a contract with redirect your traffic for you can also be problematic.

8.5 Putting CloudFront In Front of the Entire Application

With our app, the content is dynamic. This makes it difficult (but, as we will see, not impossible) to put a CDN such as CloudFront in front of it. Using CloudFront poses a problem—how do we prevent people from viewing stale content?

In fact, most CDNs can be treated much in the same way as we dealt with our local caching. We can simply set a maximum expiration date for our content.

On CloudFront, if you look at your distribution, you will see a "Behaviors" tab. Clicking this, you will see a single, default behavior. These behaviors allow you to set different settings on different content paths from your server. For this example, we only need to edit the one that is already there. Select the current behavior and then click "Edit."

If you scroll down, you will see a set of TTL (time-to-live) values, which are in seconds. To change these values, first change the "Object Caching" setting from "Use Origin Cache Headers" to "Customize." If you don't want the content to be older than 5 seconds, set the "Maximum TTL" to 5. This will mean that the cache will only keep the content around for 5 seconds before requerying for the content. Thus, the content may be a *little* stale, but not very much.

This leads to another question—how will the CDN handle sending forms to the server? This can be handled by Behavior settings as well. Under "Allowed HTTP Methods," select the option to include all of the HTTP methods. The CDN will only cache the GET and HEAD requests, but will forward all other requests straight to your server.

Additionally, by default, CloudFront doesn't forward cookies, HTTP headers, or query strings to the server. This is to decrease the number of origin requests it has to make, as well as the number of objects it has to cache. However, in our application, the content you see is based on the query string. Therefore, you need to set "Query String Forwarding and Caching" to "Forward All, Cache Based on All" to get the application to work correctly.

Once these are all set, click "Yes, Edit" to save the changes in the behavior.

You can now go to your CloudFront distribution's URL and use it as if it were the regular web site. You can also follow the instruction in Section 8.4 in order to have the user access this via your web site's own hostname.

Before you go hog wild with application setups like this, keep in mind that all cache setups are a balance. If you use too many keys for your cache (i.e., paths, headers, query strings, cookies), then most of your content will be passed through to the server (i.e., not cached), and you won't get the performance improvement. In fact, it will just add overhead and cost, because you have to pay *both* the bandwidth of the cache *and* the bandwidth at your server. However, if the cache setup is too loose, users will see out-of-date data or, worse, someone else's data! For example, if your content is personalized based on a cookie, but you don't use the cookie as part of your cache key, then if Jim Bob requests a page and it is put into the cache, when Jane Doe requests the same page, she will get Jim Bob's page!

So how does someone with a highly personalized web application make better use of the CDN? The answer to that is by turning your application inside out.

8.6 Turning Your Application Inside Out

The problem we ran into in the last section was the issue of delivering highly customized pages to users and still making good use of the cache. Having customized pages generally means that they aren't cacheable. This dilemma can be solved by employing an "inside-out" page architecture for your web site.

Historically, web applications are built by having servers generate pages. When a user navigates to a URL, the server takes a template, combines it with user data, and adds in the page content to produce the final HTML page. This HTML page is then delivered, as a whole, back to the user.

This architecture works well for a variety of reasons:

- Most web application frameworks are built around this paradigm.
- This architecture closely matches the way HTML is structured.
- This architecture is easy to conceive of and build.
- This architecture requires much less planning to implement.
- This architecture has very little up-front development required.
- Much of the Web's history is rooted in this architecture.
- Optimization of this architecture is performed last (i.e., premature optimization is the root of all evil).

The problem with this architecture, however, is that for most use cases, it is very inefficient in high-traffic situations.

An alternative way to structure your application that has been made possible through developments of Ajax techniques over the years is to turn your page inside out. That is, in a typical architecture, the server generates a page which contains dynamic content. In the new way, the server generates a page that is basically static, but *calls back* to the server for any dynamic content.

Imagine, for instance, having a page that lists your products. In typical web frameworks, your server would perform the following tasks:

1. The user's browser requests a specific page from the server.
2. The server begins web application process engine (PHP, Ruby on Rails, etc.).
3. The application accesses the database to get a list of products.
4. The application generates HTML for the product list.
5. The application accesses the database for status information on the user (i.e., login information, cart information, other status information, etc.).
6. The application generates HTML headers and footers that incorporated the user's status information and where they were in the page navigation.
7. The application stitches together the whole HTML page from the generated fragments.
8. The server sends the final HTML page back to the user.
9. The user's browser renders the page.

As you can see, this is a fairly involved process, and the user doesn't get *any* feedback until the entire page has been rendered on the server. This leads to a lot of waiting around for the server to finish up.

However, if we turn it inside out, we can achieve a much better optimization strategy. What we will do is serve up the page as a static page and then let the user's own browser be responsible for getting user-based content and stitching together the page.

The new sequence would look like this:

1. The user's browser requests the page from the CDN.
2. Assuming it is already cached, a CDN server near the user immediately responds with the page content (i.e., a list of products).
3. The user's browser loads the content and displays it immediately, with a loading spinner next to any user-specific data.
4. The page executes JavaScript, which issues a request back to the server for the user's status information.
5. The server sends back the user's JSON-encoded status information.
6. The page renders the remaining components of the web page.

In this sequence, the server's burden is greatly reduced. It no longer has to query for the product listing, generate the headers and footers, or stitch the page together. All of those happen in a very fast transaction between the user and the CDN. The server is only responsible for a user's session state (and even that can often be localized to the user's own browser!). This allows for a large amount of dynamic, interactive content, with very little burden on the server.

This concept can be expanded even further, such that even the product list is generated after the fact. API requests themselves can also be cached, either using the same or different settings as the main web pages.

As you can see, CDNs provide a very flexible and powerful tool for boosting web site speed if the application is judiciously architected.

The main drawback for this approach is the amount of planning and foresight that it takes to implement. This approach requires deciding on an Ajax and dynamic HTML framework, laying out a nice application architecture, developing APIs and API authentication, and planning the page layout that will incorporate all of the pieces. On the old architecture, you could usually just hammer out pages fairly haphazardly, and they would still work. When using the inside-out architecture, you have to start with a plan, and it requires a decent amount of up-front effort. This book does not attempt to show code for what an inside-out page architecture looks like, precisely because it requires a lot of code to implement.

In any case, no matter what your application architecture, most applications can benefit from some sort of CDN solution.

CHAPTER 9

Using S3 for Infinite Disk Space

For a truly scalable site, another aspect that is often needed is unlimited disk storage. Managing large-scale storage for servers is truly a difficult feat. Deciding on how much redundancy, how much accessibility, how many disks per server, and how to manage the disks to make sure you know ahead of time if a disk is failing is a tough task. Even for small-scale sites, managing files can be difficult.

Thankfully, using a file storage service will allow you to outsource these tasks to a third party, probably for a much smaller price than if you tried to do it yourself. The gold standard for file storage services is Amazon's S3 service. S3 stands for Simple Storage Service. This acronym is largely true—S3 is fairly easy to set up for the simple cases, but it also has quite a bit of flexibility in there for more complicated tasks.

S3 gives you an infinite amount of space at a very low cost per gigabyte. It will scale with you and prevent you from all of the file management headaches that come from storing files locally.

Other cloud storage services exist as well, many with better pricing structures. Some of the more common ones include Backblaze B2, DigitalOcean's Spaces, and Rackspace's Cloud Files. Here we use S3 because it has the most widespread adoption and integration.

J. Bartlett, *Building Scalable PHP Web Applications Using the Cloud*,
https://doi.org/10.1007/978-1-4842-5212-3_9

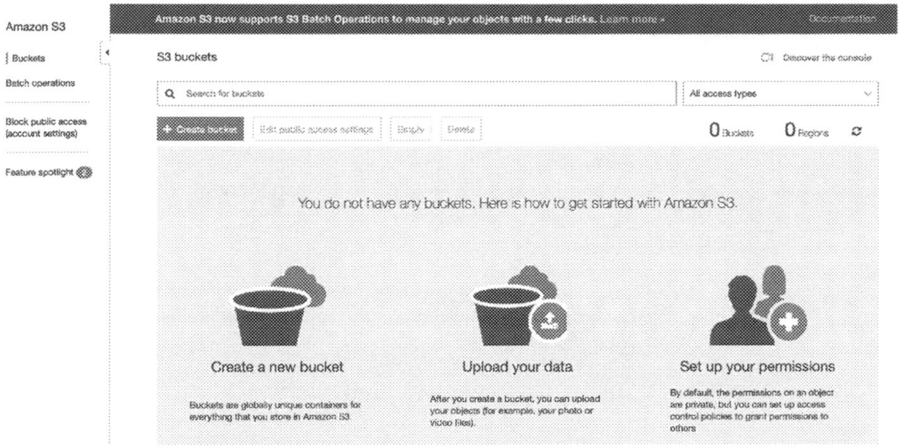

Figure 9-1. *The Initial S3 Screen*

9.1 Getting Started with S3

S3 is part of the Amazon AWS suite of tools. Therefore, you can use the same login that you created in Chapter 8 to access AWS.

Once you sign in to AWS (the URL is `http://aws.amazon.com`), you can access S3 under the "Storage" heading. When you click S3, you will get a screen similar to Figure 9-1.

The main button for this screen is the "Create Bucket" button. S3 organizes all of its files into what they call "buckets." A bucket is kind of like a named hard drive. That's where you store all of your files. Bucket names have to be unique, not only to your account but actually across all of Amazon. Therefore, you should not rely on any particular naming convention for Amazon buckets which presume that you can predict what names will be available in the future. Instead, it is better to have bucket names be configurable in your application so that it is easier to manage.

When you click "Create Bucket," it will ask you for a bucket name and a region. AWS organizes almost all of its services except CloudFront into regions. For our purposes, the region itself does not make a lot of difference. But, if you have a specific reason for needing a bucket in a specific physical location, AWS allows you to choose where it goes. Click the "Create" button to create your bucket.

Once you successfully create your bucket, you should get a list of your buckets (with just your one bucket). If you click your bucket, you can browse your empty bucket. To get a feel for S3, just go ahead and upload a file from your hard drive to the bucket. Click the "Upload" button, and you can then drag and drop files from your computer up to your bucket. Click "Upload" to get them uploading.

9.2 Folders in S3

If you look around on the S3 console, you will notice that you have the ability to create folders/directories in your buckets. However, in S3, folders are not actually real. S3 buckets in reality have a completely flat structure with just filenames and files (technically, the filenames are called "keys" and the files themselves are called "objects"). However, the filename can contain a slash character. The AWS user interface then uses the slash characters to show you the files as if they are in folders. When you "create a folder" in S3, it is actually creating an empty file of the folder's name, with the name ending in a slash. In short, the S3 console makes it look like you have folders and subfolders, but really it is just a giant wad of files, some of them having slashes in their names that the S3 console uses to separate out into fake folders to make it easier to look through.

9.3 Getting Credentials

Before we connect our S3 account to our server, we need to create a set of security credentials. For this, AWS uses a system known as IAM, or "Identity and Access Management." IAM allows you to create users that have restricted privileges so that if your security keys get compromised, it does not let an attacker take total control of your environment. As with other services, IAM can be found by searching under their service list for IAM. When you first load the IAM screen, it will look like Figure 9-2.

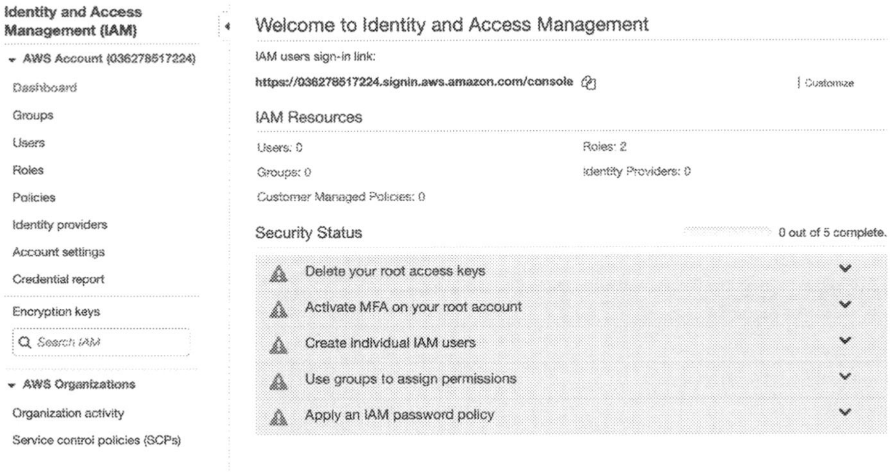

Figure 9-2. *IAM Initial Dashboard*

IAM manages permissions mostly using "Groups," which are essentially containers for permissions. Therefore, we will start by creating a group. Start by clicking "Groups" on the left panel, which will bring up an empty group listing. Then Click "Create New Group." This brings up a screen that asks for a name for the group. We will use the name `guestbook-access` (the name doesn't matter, we will just have to reference that name later). Click "Next Step" to continue.

Next you will attach policies to the group. Policies are complicated permission groups. Thankfully, AWS has very helpful predefined policies. For our purposes, we only need the policy called "AmazonS3FullAccess." You can search for it in the filter box and then select it when you find it. Figure 9-3 shows what this looks like.

Finally, it will ask you to review and finalize your group. Click "Create Group" to finish.

Now you can add a user to the group. On the left-hand side of the screen, click the "Users" link. This will bring you to an empty list of IAM users. To get started, click the "Add User" button.

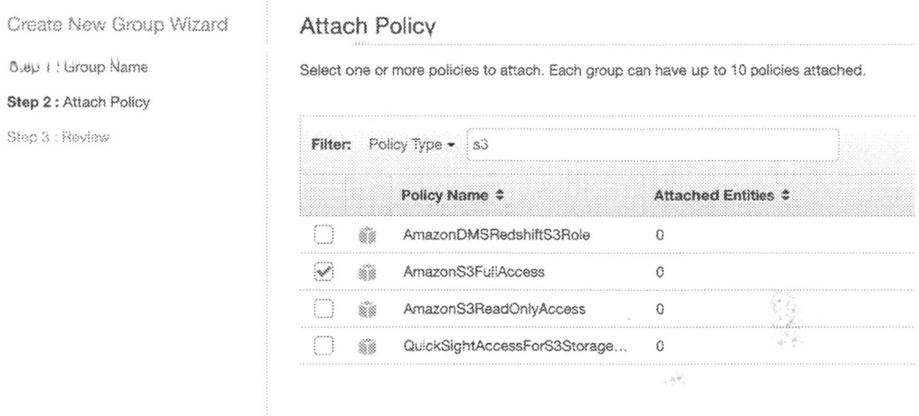

***Figure 9-3.** Attaching a Policy to a Group*

In the next screen, it will ask for the user name and the access type, as shown in Figure 9-4. We will call the user `application-user`, though the actual name doesn't really matter. Under "Access Type," select "Programmatic Access." This means that the created user won't be able to log in, but will only be able to use the API.

In the next screen, it will ask you to add the user to a group. Simply select the group that you previously created (we called ours `guestbook-access`). The next screen allows you to set up tags for this user. We don't need any, so you can continue on past this screen. Finally, it will ask you to review your information. At this point, you can click "Create User," and it will create the user for you.

After your user is created, you can now download their credentials. The screen looks like Figure 9-5. It lists the user and then two special fields: "Access Key ID" and "Secret Access Key." These two fields essentially operate as a resettable username ("Access Key ID") and password ("Secret Access Key") for the API for this user. You can either download the credentials or copy them from the fields on the screen.

Going forward, we will refer to the actual Access Key ID as `MYACCESSKEYID` and the actual Secret Access Key as `MYSECRETACCESSKEY`.

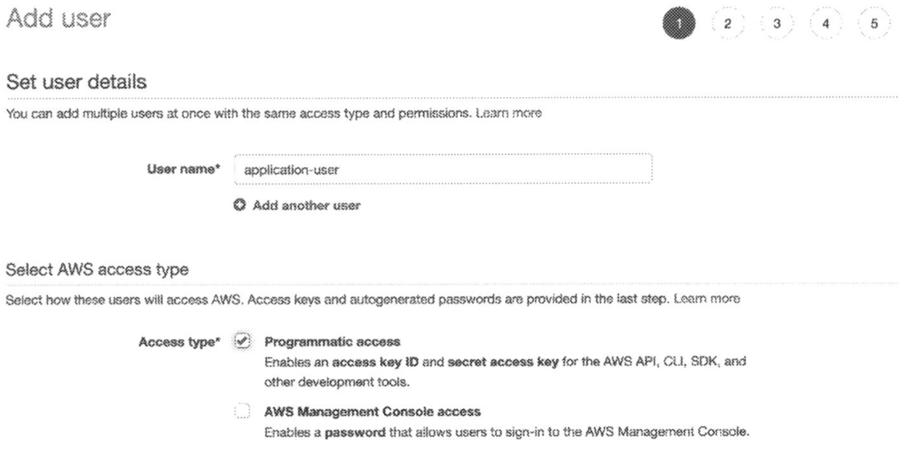

Figure 9-4. *Creating an IAM User*

If you later lose those credentials, you won't be able to obtain them again. However, you can go back into the user's record and create a *new* set of credentials. If, at a later time, your server's security is compromised, you can deactivate the old credentials and issue new ones for the same user.

9.4 Access S3 via Command Line

AWS has a command-line tool that allows access to not only S3 but a wide range of their scalability APIs. To install this, issue the following command as root on the template node:

```
yum install -y awscli
```

The AWS command line has two main ways to specify your Access Key ID and Secret Access Key. You can either do it via environment variables or via a configuration file. Environment variables are more flexible, so we will do that route.

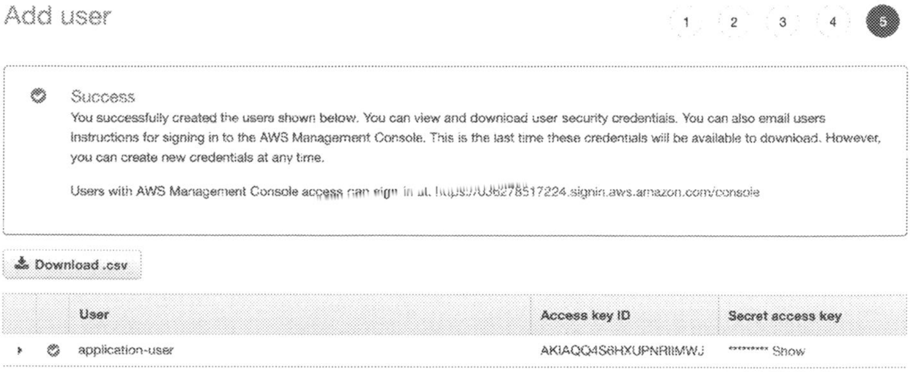

Figure 9-5. *Retrieving IAM Credentials*

If you are not familiar with the command line, an environment variable is a variable that is set in your command-line session. Not only that, but commands that you call have access to all of your environment variables. Additionally, any environment variables that you set go away when you log out, so they will need to be reset every time you log in (if you want them to be set automatically upon login, you can put the commands to set them in a file called `.bash_profile`).

The environment variables that the `aws` command uses for your credentials are `AWS_ACCESS_KEY_ID` and `AWS_SECRET_ACCESS_KEY`. To set these variables, enter the following commands in your terminal (replacing `MYACCESSKEYID` and `MYSECRETACCESSKEY` with your actual keys):

```
export AWS_ACCESS_KEY_ID=MYACCESSKEYID
export AWS_SECRET_ACCESS_KEY=MYSECRETACCESSKEY
```

Now you can use the `aws` command to manipulate your S3 buckets. To get a list of your buckets, issue the following command:

```
aws s3 ls s3://
```

It should list the bucket that you created in Section 9.1.

To list the contents of that bucket, issue the following command (replace BUCKET with the actual name of your bucket):

```
aws s3 ls s3://BUCKET/
```

To understand the way the command works, `aws` is the main command we are using, `s3` tells which group of subcommands to work with, `ls` is just like the `ls` command on Linux (it lists the contents), and `s3://BUCKET/` is the location that we would like to look at.

The `aws` command also gives other common commands for manipulating files. Instead of `ls`, we can use `cp` to copy files in and out of the bucket. If you had a file called `test.txt`, you could copy it to your bucket with the command:

```
aws s3 cp test.txt s3://BUCKET/
```

To copy a file *from* your bucket *to* your node, just switch the places of `s3://BUCKET/` and `test.txt`.

Additionally, you can create temporary access URLs for your files. These URLs are signed URLs. This means that AWS knows that an authorized person generated the URL, and AWS will trust that URL for a specified amount of time as a valid means of accessing that file.

This allows you to direct people directly to the AWS site to retrieve data that they need, instead of having to transfer it to your server and then send it yourself. It saves processing power, bandwidth, and response time.

However, in order to do this, we will need to know what region the bucket is in. You specified a region when you created the bucket, but AWS doesn't always show the "computerized" version of the region, which you will need for the command line. Issue the following command to find the region of your bucket (replacing BUCKET with your bucket name):

```
aws s3api get-bucket-location --bucket BUCKET
```

This will return a JSON-encoded value. The key is called `LocationConstraint`, and the value is the name of the region that the bucket is hosted in. Some common values for the region are strings like us-east-1, us-east-2, ca-central-1, eu-west-2, and others. We will use REGION to represent whatever your region is.

To get a URL to access your file, use the following command:

```
aws s3 presign s3://BUCKET/FILE --region REGION
```

This will generate a URL that you can copy and paste into your browser. Lo and behold—the file will appear! However, this URL will only be good for 3600 seconds (1 hour). If you want the URL to be good for a different amount of time, you can tell it using the --expires-in flag. So, if you want the URL to expire in 20 seconds, you would just add --expires-in 20 to the command.

SPECIFYING ENVIRONMENT VARIABLES INLINE

Before we start coding, I wanted to take a quick sidebar and let you know another way to set environment variables. You can set an environment variable so that it will be valid only for one application by specifying the environment variables to set on the same line as a given command before the command itself.

So, for instance, if I were to run the command example-command, and I wanted to set the environment variable MYVAR to myval, I could run the command like this:

```
MYVAR=myval example-command
```

This would set the environment variable, but only for running the command. You can actually set as many environment variables as you wish when running commands, they just have to be separated by spaces. So, for instance, to set two values, we could do:

```
MYVAR1=val1 MYVAR2=val2 example-command
```

This is how we will be setting our credentials in the application we will create. For configuration purposes, it is actually best to leave the credentials (and other configuration information) out of your code and set them all via environment variables on the server. However, that requires more in-depth, server-specific configuration details than is appropriate for this book.

9.5 Connecting Your Application to S3

Now that we know how to talk to S3 from our server, we will now connect our guestbook application to S3 so that users can upload an image with their message. This is actually fairly straightforward. All we have to do is

1. Create some common functions to grab AWS configuration information.
2. Allow our form to have an image field.
3. Check for an image upload when a guestbook entry is created.
4. Transfer the image to S3.
5. Create a signed S3 URL for the image when viewing a guestbook entry.

The first step is creating some helper functions in `common.php` for handling AWS credentials and configuration. Figure 9-6 shows the functions for that. Remember to replace BUCKET, ACCESSKEY, and SECRETKEY with your own values for these items.

```
function getS3BucketName() {
  return "BUCKET";
}

function getS3Region() {
  return "REGION";
}

function getAWSCredentials() {
  $s3ak = "MYACCESSKEYID";
  $s3sk = "MYSECRETACCESSKEY";
  $s3ak_env = "AWS_ACCESS_KEY_ID=$s3ak";
  $s3sk_env = "AWS_SECRET_ACCESS_KEY=$s3sk";
  $creds = "$s3ak_env $s3sk_env";
  return $creds;
}
```

Figure 9-6. *Additions to `common.php` for AWS Configuration*

The function `getAWSCredentials()` will return the credentials as a string which can set the environment variables when prepended to a command string.

The next step is putting an image field on the form in `new.php`. There are two parts to this. The first thing we must do is modify the `<form>` tag so that it will allow a file upload. To do that, add the attribute `enctype="multipart/form-data"` to the `<form>` tag. Without that, the file input tags will not actually upload files.

Next, we need to add a file upload field. Right before the submit button, add the following lines:

```
<label>Image (JPEG)</label>
<input type="file" name="imagefile" />
```

Now your form is configured to have a file upload. Next, we will configure `create.php` to accept the file upload.

```
$tmpname = $_FILES["imagefile"]["tmp_name"];
if($tmpname) {
  $has_img = true;
} else {
  $has_img = false;
}
```

Figure 9-7. *Modification to `create`.php to Detect File Upload*

There are two parts we have to change. First, replace the line `$has_img = false;` with the code in Figure 9-7. This code detects if a file was uploaded and, if so, makes sure the database is updated to reflect that. Second, add the code in Figure 9-8 immediately after the line that says `$stmt->execute();`. This is the code that actually transmits the file to S3.

```
if($has_img) {
  $last_id = $dbh->lastInsertId(
    "gb_entries_id_seq"
  );

  $s3_creds = getAWSCredentials();
  $s3_base = "s3://" . getS3BucketName() . "/";
  $s3_file = $s3_base . $last_id . ".jpg";
  $s3_creds = getAWSCredentials();
  $cmd="$s3_creds aws s3 cp $tmpname $s3_file";
  exec($cmd);
}
```

Figure 9-8. *Modification to create.php to Transmit File to S3*

Now we just need to provide a way to view the image if you click the guestbook entry. To do this, we need to modify `single.php`. Just add the code in Figure 9-9 right before the line that says `getFooter()`.

```
if($row["has_img"]) {
  $s3_url = "s3://" . getS3BucketName() .
            "/" . $row["id"] . ".jpg";
  $s3_region = getS3Region();
  $s3_creds = getAWSCredentials();
  $seconds = 300;

  $cmd = "$s3_creds aws s3 presign $s3_url" .
         " --region $s3_region" .
         " --expires-in $seconds";
  exec($cmd, $cmd_output);
  $url = $cmd_output[0];
?>
  <label>Image</label>
  <img src="<?php echo h($url) ?>" />
<?php
}
```

Figure 9-9. *Modification to* `single.php` *to Show Image*

This code will request a signed URL for the upload and then put it in an image tag for viewing.

Note that for a real application, we would want to validate that the uploaded file was indeed a JPEG file. Otherwise, anybody could upload anything, and hackers could easily abuse the system as a free file-sharing site, or for other nefarious purposes. Additionally, you would probably want to have an administrative function which verifies that the images uploaded are appropriate. Otherwise, someone could easily turn your guestbook into a porn-sharing site. Those are outside the scope of our simple example application, but they are good things to keep in mind.

Also remember that you will wind up paying for not only storage space but also usage bandwidth for all traffic in and out of S3. Failure to be vigilant can wind up being costly.

Now you need to test out your new code on `template_node`, and then, when it is working, deploy it to your cloud cluster by reimaging the servers as described in Section 6.3.

Now try out your application—it now has infinite file storage with Amazon S3!

Note that there is also an AWS library for PHP which has S3 functionality with it. Here, we used the command line since we already learned that tool in the preceding section. Information about the AWS PHP library is available at `https://docs.aws.amazon.com/sdk-for-php/`.

☞ A NOTE ON S3 SIGNATURE EXPIRATION

One thing to keep in mind when dealing with signed URL is to be wary of how they interact with caching. In this case, `single.php` is not cached, so there is no real worry. However, if it was, it is important to make sure that the expiration time of the URL is significantly longer than the amount of time that the rendered code might sit in the cache.

For instance, if the URL was only valid for 30 seconds, but the cache lasts an hour, then, after the first 30 seconds is over, for the rest of the hour users will be getting URLs that they cannot make use of. Just keep this in mind when (or, preferably, before) problems start arising.

☞ S3 FILE PERMISSIONS

In addition to signed URLs, it is also possible to allow access to S3 files through granting public access to the files. This works, but it isn't the best approach to sharing files. The problem is that if you just provide a publicly shared URL for people to access, then this URL could be shared around, and people could just use your S3 resources for their own purposes, bypassing your web application altogether. This means that you might wind up paying to be someone else's file server.

Even if you don't implement careful controls over file access in your application immediately, forcing everyone to use signed URLs controlled by your application to access S3 objects (as we have shown here) means that when you are ready to implement access controls, everything is ready to do so. At minimum, it prevents the simple sharing of S3 URLs publicly on the Internet which use your AWS account resources for large file downloads.

If you really prefer doing this over signed URLs, you can manage this by first enabling public access on the bucket itself and then setting individual files to be readable by everyone. It also means that you will need to set permissions on newly uploaded files. You can do this on the `aws s3 cp` command by adding the following flag to the end of the command (the URL should be on the same line as the rest of the line, with no space after either equal sign):

```
--grants read=uri=http://acs.amazonaws.com/groups/global/
AllUsers
```

However, this only works if the bucket itself has been configured to allow public access.

CHAPTER 10

Hosting with AWS

While most of this book has focused on hosting your application on Linode, since so much of the cloud hosting is based on AWS, I think it is worthwhile to look at some of the hosting options on AWS.

One of the issues with AWS is simply its massive number of options. The number of options available actually makes it quite difficult to administer.

As an example of this, one time I was working on a team and needed access to the logs. After much searching, we finally found the option to grant access to the log files. However, it turned out that *looking at* the log files was actually a *different* permission than the permission needed to *retrieve* the log files. So, while I had permission to *look* at the log files, I didn't have permission to actually get them. They finally gave up fine-grained control and made me an administrator. That's not to say that AWS doesn't have this ability, it's just that trying to manage it turns out to cost as much in time as effort as you might have saved doing it some other, less "automated" way. Having lots of options and trying to be everything to everyone can easily make it so that you drown in options and settings that no one has the time and opportunity to master.

There are so many fine-grained controls, each with their own issues, quirks, and bugs. The number of different platforms and systems that AWS "supports" is very large. However, I find that oftentimes there are large holes in that support, which, while they can be worked around, the workarounds sometimes make me wish that I had just done it all manually to begin with.

J. Bartlett, *Building Scalable PHP Web Applications Using the Cloud*,
https://doi.org/10.1007/978-1-4842-5212-3_10

10.1 Using Amazon Lightsail

Amazon's first foray into cloud hosting is called EC2—the Elastic Compute Cloud. Like most of AWS, this is a very flexible option for cloud hosting. However, it suffers from the problem inherent in a lot of AWS—it is really complicated to set up and use successfully. Not only is the setup complicated, the *pricing* is ridiculously complicated. They charge you not only for computer time but also for hardware I/O requests. That's right, they keep track *and bill you* for the number of hard drive accesses.

In order to better compete with easier-to-use services like Linode and DigitalOcean, AWS introduced Amazon Lightsail. Lightsail's feature set and pricing structure is very similar to what we have looked at on Linode. Lightsail is accessible from the same AWS management console that contains CloudFront and S3. Just search for Lightsail, and the dashboard should look similar to Figure 10-1.

There are a few minor differences between Lightsail and Linode to be aware of for our purposes:

1. Lightsail instances by default connect via a public key instead of a password, and connect you to a regular user instead of connecting as root.

2. Lightsail's CentOS distribution has a slightly different set of preinstalled packages than Linode.

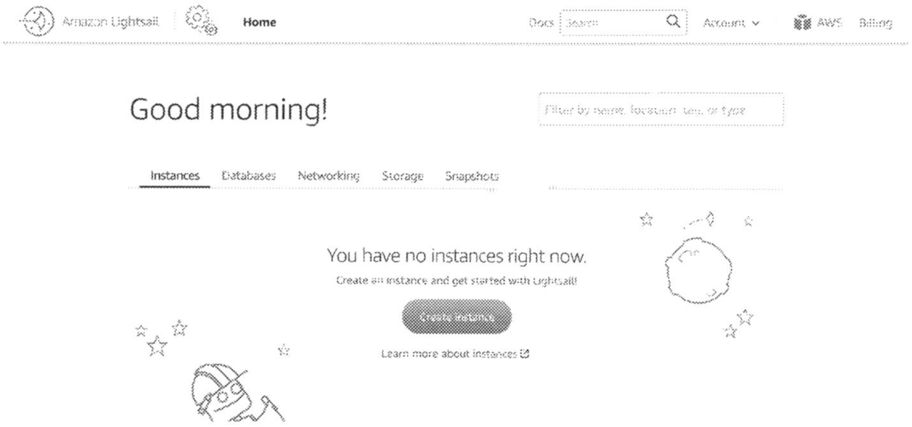

***Figure 10-1.** The Initial Dashboard of Amazon Lightsail*

3. All Lightsail instances are automatically added to Amazon's private network. In fact, when you are on the box, the *only* IP address you see is the private network address. The external IP that you receive just gets routed to that internal IP address.

4. Backups of Lightsail instances have to be made manually.

5. Lightsail has a separate database service that you can use if you don't want to manage your database server yourself.

To launch and set up a new instance, click the "Create Instance" button. This brings you to a screen similar to Figure 10-2.

Here, you will need to choose the datacenter (instance location) which will host your node. This is slightly different than Linode, as AWS has both datacenters and *availability zones*. In AWS, datacenters are organized into multiple availability zones. These zones can easily share resources for load balancing, but each availability zone has a separate line for power and external Internet connections. Essentially, they act like they are in the same datacenter, but servers in different availability zones are unlikely to be affected by the same event (such as loss of power or Internet). So, for instance, you could have your master database in one availability zone, and a replica database in another. Then, if the availability zone of your master goes down, the replica (since it is in a different availability zone) can be promoted and serve as the master.

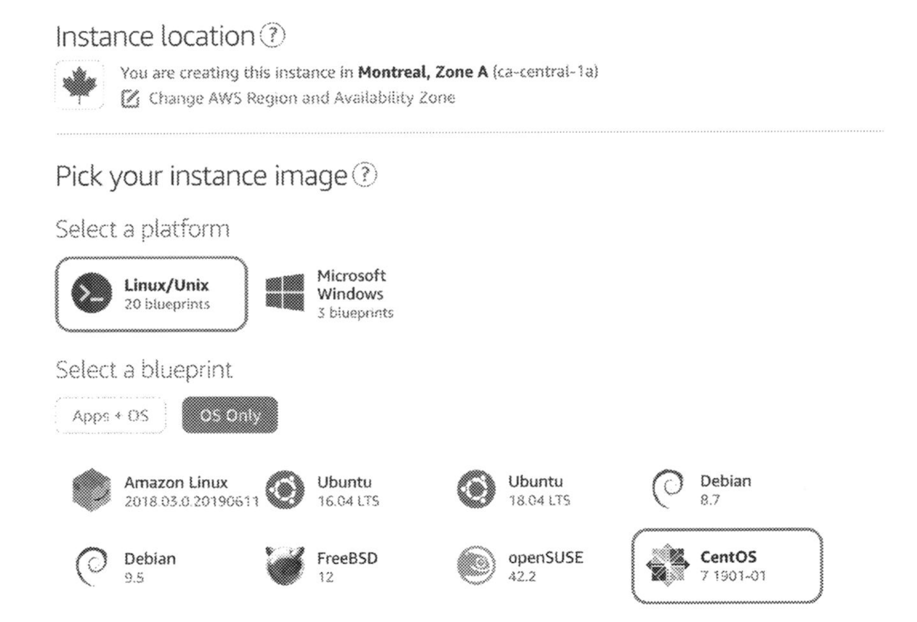

***Figure 10-2.** Creating a Lightsail Instance*

The platform we want to choose is “Linux/Unix.” Then, select “OS Only” (the other options allow for preconfigured servers for specific tasks). After that, select “CentOS.”

Scrolling down, you can ignore most of the other questions, and select whichever size of node you wish. Then, to identify your instance, call it `template-node` as we did before. Click “Create Instance” to create the machine.

It may take AWS a while to create your machine. When it does, you can click the machine, and you will get a dashboard similar to Figure 10-3.

To log in, just click the button labeled "Connect Using SSH." This will bring up a terminal window, with you connected to the node as the user `centos`.

You can use this `centos` user in place of the `fred` user in the examples in this book. Or, you can go ahead and create `fred` when requested. You will still need to use the box as `root` to configure it, however. To log in as `root`, just issue the command:

```
sudo su -
```

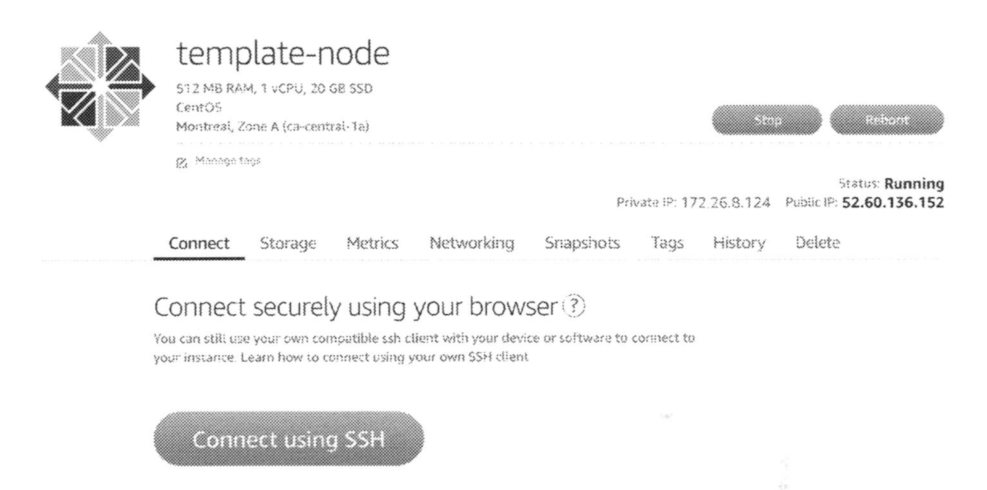

Figure 10-3. *Lightsail Node Dashboard*

And now your session will be as the `root` user.

Lightsail nodes come with a slightly different set of packages than Linode nodes. To get your Lightsail node similar to the starting point of Linode's system, issue the following commands:

```
yum install -y nano
yum install -y firewalld
systemctl start firewalld
systemctl enable firewalld
```

From this point, you can follow the node setup instructions from Chapters 3 and 4 essentially identically.

In Chapter 5, the mechanism for creating new instances from backups is only slightly different. In Lightsail, backups are created using the "Snapshots" feature. All you have to do is go to the "Snapshots" tab of your node, give your snapshot a name, and create it. Once created, you can create a new node from this snapshot as shown in Figure 10-4. Load balancers can be created from Lightsail under the "Networking" tab off of the main screen.

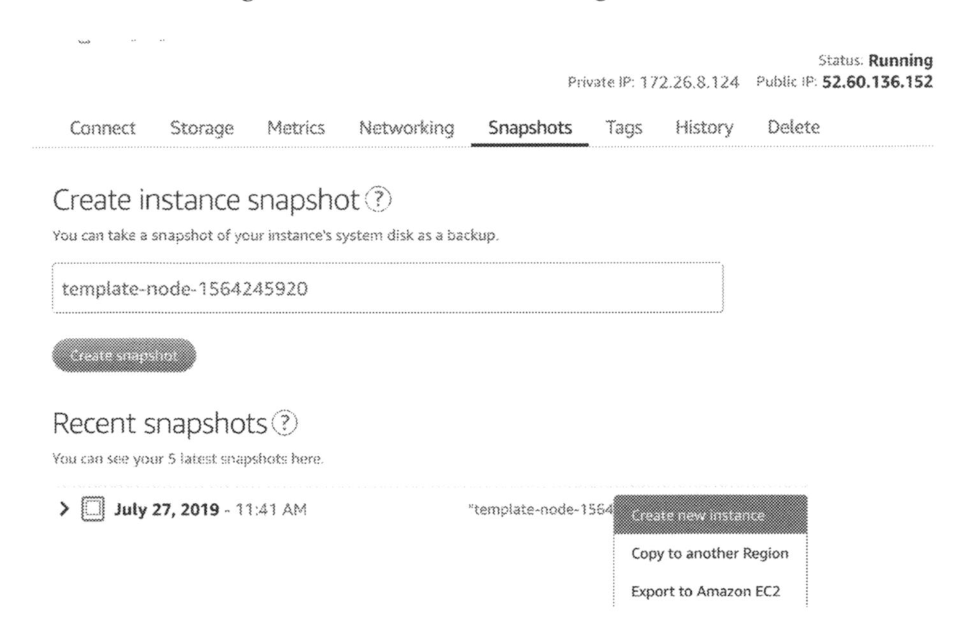

Figure 10-4. *Creating a Node from a Snapshot*

After that, the rest of the information about creating a cloud doesn't change, as it is all about what happens on the nodes themselves.

10.2 Hosting on Elastic Beanstalk

While most of this book has been focused on Infrastructure-as-a-Service (IaaS) clouds, I did want to spend at least some time introducing a Platform-as-a-Service (PaaS) cloud. AWS has a PaaS cloud called "Elastic Beanstalk" which runs a variety of different application types, one of which is PHP.

A PaaS cloud takes away all of the system administration work from cloud computing. The problem, though, is that most advanced cloud systems wind up requiring some amount of system administration anyway. It's not that a PaaS system is unusable when you have strange configuration requirements, but rather the amount of effort to configure your PaaS system correctly, maintainably, and in sync with what your platform provider is also doing winds up being more than if you had just taken full control like you do with an IaaS cloud. In any case, in this section, we will go over what it takes to get our application up and running with Elastic Beanstalk.

In a PaaS system, we have no control over the machines. Therefore, we can't designate a particular machine as a database server, a job server, and so on. Instead, each application can be scaled up or down across as many machines as the PaaS system wishes. This generally means that the PaaS system will manage the database.

For AWS, this means using their relational database service—RDS. Connecting the application to RDS is pretty straightforward. Elastic Beanstalk will set up environment variables for all of the connection information, and we just have to program our app to read them.

In order to do this, we just have to change `common.php` and replace the `getReadOnlyConnection()` and `getReadWriteConnection()` functions. Both of those functions should look like Figure 10-5.

```
return new PDO(
  "pgsql:host=" . getenv("RDS_HOSTNAME") .
  ";port=" . getenv("RDS_PORT") .
  ";dbname=" . getenv("RDS_DB_NAME") .
  ";user=" . getenv("RDS_USERNAME") .
  ";password=" . getenv("RDS_PASSWORD")
);
```

Figure 10-5. *Accessing a Database with RDS*

This will put together a connection string based on the environment variables that RDS sends in.

Second, since you will not have direct access to the database, we need a PHP script which will create the database tables for us. Create a file called `createdb.php` that has the following code in it.

```
<?php
        include("common.php");
        $dbh = getReadWriteConnection();
        $stmt = $dbh->prepare(
    "create table gb_entries " .
    "(id serial primary key, name text, " .
    "email text, message text, " .
    "created_at timestamp, " .
    "has_img bool default false)"
  );
        $stmt->execute();
?>
```

Figure 10-6. *Creating a Database with RDS*

Note that, since we don't have a specific server, these changes are being made to a local copy of the application on our personal computer's hard drive. Once those changes have been made, you are ready to get started.

To start with Elastic Beanstalk (EB), go back to the AWS management console and search for "Elastic Beanstalk." On the dashboard, click the button titled "Create New Application." Give it a name, and click "Create."

On EB, applications can be compartmentalized into "environments." Environments can be used for all sorts of things, including having staging vs. production environments, having different groups of servers for different tasks, and having fast switching between different versions of your application. For our purposes, we will only have one environment. Therefore, click the button to create a new environment. It will ask you what type of environment to create. We want a "Web Server Environment."

Create a web server environment

Launch an environment with a sample application or your own code. By creating an environment, you allow AWS resources and permissions on your behalf. Learn more

Environment information

Choose the name, subdomain, and description for your environment. These cannot be changed later.

Application name example-application

Environment name ExampleApplication-env

Domain Leave blank for autogenerated value .us-west-2.elasticbeanstalk.com

Description

Base configuration

Platform Preconfigured platform

Platforms published and maintained by AWS Elastic Beanstalk.

PHP

Custom platform

Platforms created and owned by you. Learn more

-- Choose a custom platform --

Application code Sample application

Get started right away with sample code.

Figure 10-7. *Creating a New Environment*

The next screen is a configuration screen, as shown in Figure 10-7. You can name your environment if you want, but the only important setting on that screen is to select a "Preconfigured Platform." We, obviously, want "PHP." By default, it will load a sample application. That is what we want for now. Click "Create Environment" to finish the process.

Once created, you will have a dashboard that looks like Figure 10-8. In the "Overview" section, it gives the basic health of the app, the version, and the platform. Under "Recent Events," it lists all of the actions that the system has recently undergone. Note that *everything* in EB winds up taking a lot longer than you might think, but "Recent Events" helps you keep tabs on it and gives you something to watch while you wait. On the left side, the three areas we will concentrate on are "Dashboard" (where we are now), "Configuration" (how the app is set up), and "Logs."

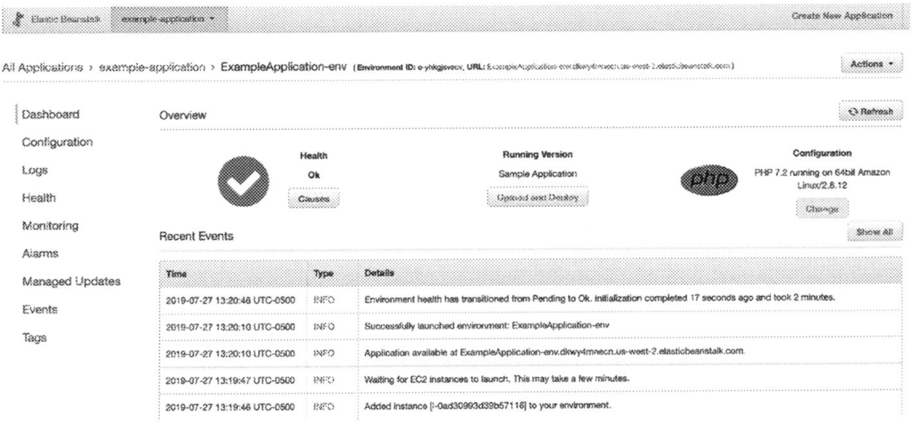

Figure 10-8. *EB Environment Dashboard*

Since we started with a sample application, you can already load up the sample application in your web browser. On the dashboard, it has a link to your application's URL. You can click it, and it will take you to the sample application.

Before we upload our application, we need to first set up the database. To do this, go to the "Configuration" section, and then go to the "Database" and click "Modify." This will allow you to create an RDS instance for your application. Set the "Engine" to be `postgres`, give it a username and password (be sure to write these down!), and then click "Apply." When the health status comes back as a checkmark (which can take 10–20 minutes), you are ready to upload your application.

To upload your application, zip up all of your files together into a single zip archive. Then, on the dashboard, click the button that says "Upload and Deploy." Choose the zip file from your hard drive, give the version a name (the specific name doesn't matter), and then click "Deploy." The health status will change to a refresh spinner, and it will deploy your application. When the spinner changes back to a checkmark, your application is deployed!

However, we're not quite done yet. When you click the link, it will give you a "Forbidden" message. This is because we don't have an `index.php` file (though you can add one if you wish). However, we need to create our database, so we need to first navigate to the `createdb.php` file to create the database (just add `/createdb.php` to the end of the URL in your browser and hit enter). This should bring you a blank page, which is fine.

Now, if you change the URL to go to `list.php`, everything should be in working order!

There are now many things that you can do with your application to make it scalable. These are all available from the "Configuration" tab of your environment. By default, EB creates environments that are single server. To upgrade it to a load-balanced application, just go into the "Capacity" section and change the "Environment Type" to "Load Balanced." This will give you a lot of options that you can play with. The most simple (and most important) is the "minimum" and "maximum" number of instances. Set the minimum to 2 to make sure that it starts up at least two machines for you. Click "Apply," wait a few minutes, and your application is now load balanced.

Other changes you can make include

- Under "Instances," you can change the size of each individual instance.
- Under "Load Balancer," you can change a lot of variables, including the balancing mechanisms.
- Under "Rolling Updates and Deployments," you can change the deployment policy to eliminate downtime during updates (the "Immutable" option is the best for production, but it takes a long time to perform updates).
- Under "Software," you can set environment variables and web server configuration options.
- Under "Database," you can change the size of your database. You can also click the "Endpoint" link to take you to the RDS management console for additional database management utilities.

After making configuration changes, hitting "Apply" will redeploy your application with the given changes, which may take several minutes. If you have any problems with your running application, you can go to the "Logs" section and download and view the most recent log messages.

CHAPTER 11

Using the Google Cloud Platform

Now that you have experience with multiple cloud providers, I hope that you are realizing that the basic components of clouds don't really change all that much from provider to provider. Each one may call something different, may have a little different process for setting something up, or may have a few additional abilities or limitations compared to someone else. However, at their core, the basics of IaaS cloud hosting are pretty similar from provider to provider.

This is itself an advantage of IaaS. It means that if you want to migrate from one IaaS to another, it is actually a fairly straightforward process. It may take planning and execution time, but, since the nodes themselves are under your control, you can make the environments pretty identical to each other.

The next provider we are going to take a look at is the Google Cloud Platform (GCP). GCP is the new kid on the block, having been established in 2008 (Linode started in 2003, and AWS came online in 2006). GCP is very similar to AWS, although it is a bit more cumbersome to configure. You can get started with GCP by signing up at `https://cloud.google.com`.

J. Bartlett, *Building Scalable PHP Web Applications Using the Cloud*,
https://doi.org/10.1007/978-1-4842-5212-3_11

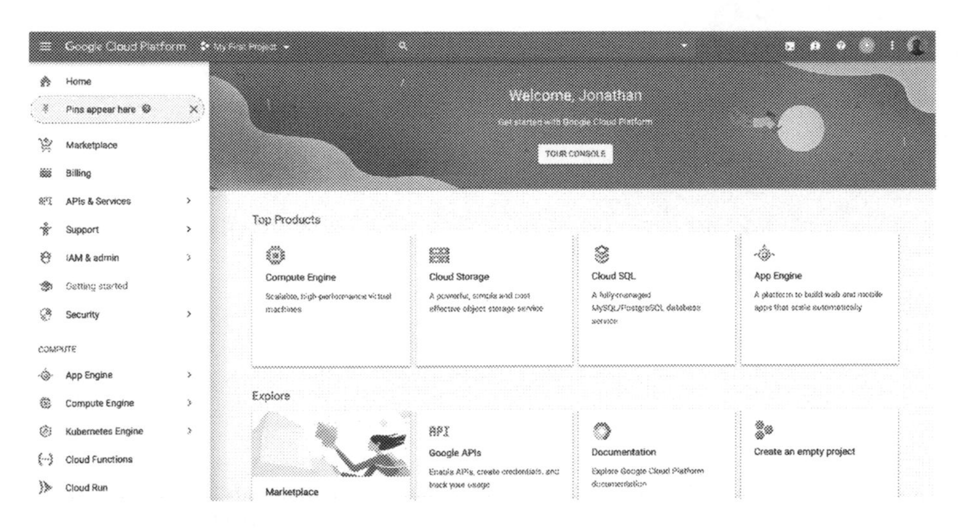

Figure 11-1. *The GCP Welcome Screen*

11.1 Setting Up Your Template Node

The first thing to know about GCP is that everything is organized into “Projects.” Each project is kind of like its own account, each with their own resources, services, and so on, but all accessible within the same login.

When you log in to the Google Cloud Platform, it should look something like Figure 11-1. Notice on the top bar, next to the text “Google Cloud Platform,” it lists the name of your current project. You can click the current project to change projects or create a new one. For this example, I am going to create a new project called `Book Examples Project`.

The left-hand side of the screen lists the GCP services. To create a new machine, go to the “Compute” section of the services, click “Compute Engine,” and then click “VM Instances.” GCP calls its machines (or nodes, as Linode calls them) “VM Instances.” To create a new machine, click the “Create Instance” button. That will bring up a screen similar to Figure 11-2

We will call our machine `template-node` and choose the smallest machine type ("f1-micro" in this case). For the boot disk, we will choose CentOS 7. As you scroll down, there is a "Firewall" section. Make sure that "Allow HTTP Traffic" and "Allow HTTPS Traffic" are both enabled. The rest of the defaults you can leave alone. When everything is set, click "Create" and GCP will create a new machine for you, and return you to the VM Listing screen.

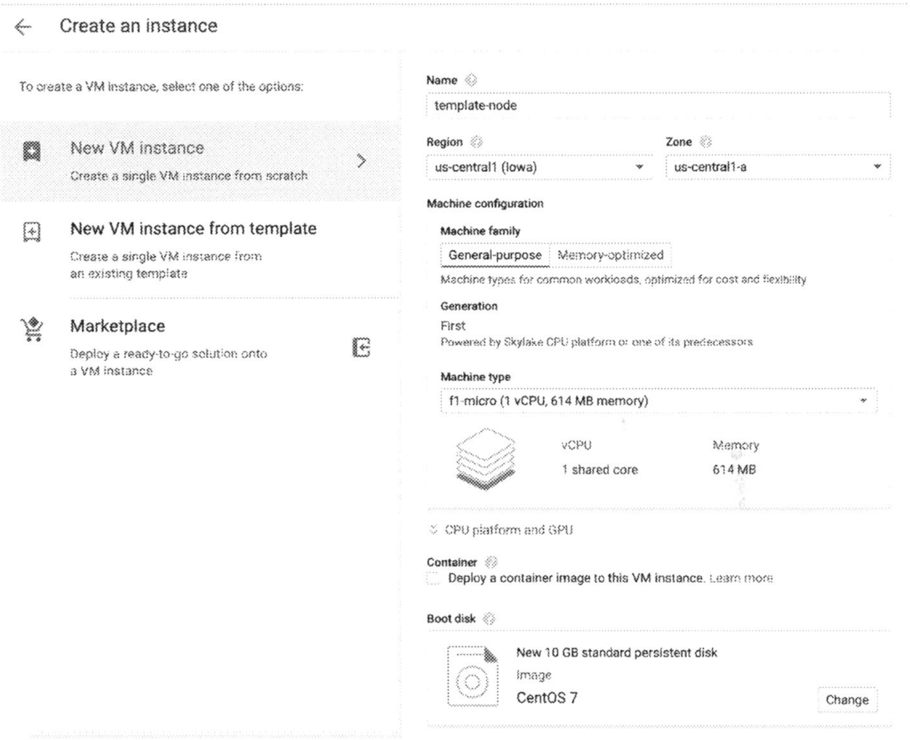

Figure 11-2. Creating a GCP VM Instance

When the machine is created, you can log in using the "Connect" option. In the drop-down menu in the "Connect" column, choose "Open in Browser Window," and it will give you an `ssh` session in your browser.

This installation of CentOS is very similar to the Linode installation, except that (a) it doesn't include nano (which you can fix with a simple `yum install -y nano`), and (b) it autocreates a user for you and logs you in as that user. You can easily switch to root by running the command:

```
sudo su -
```

From that point, you can install nano and perform all of the configuration steps outlined in Chapters 3 and 4 essentially identically.

PUBLIC VS. PRIVATE IP ADDRESSES

One interesting difference between GCP and Linode is that, on Linode, your node's public IP address is physically attached to your node. That is, when you issue `ip addr show`, it shows the public IP address. When you add a private IP address, it adds that private IP address to your node.

However, in GCP, your node *starts out* with both a private address and a public address. However, the private address is the only one that is physically mapped to the device. The public IP address is configured on the networking equipment to forward those requests to your machine.

Therefore, on all your configurations on your machine, you will use the private IP address that comes on your box. Additionally, you don't have to worry about only listening on your private IP address, because that's really all you have anyway. The GCP networking controls what services can get to your box from outside (this is why you checked the HTTP and HTTPS boxes when setting up your VM Instances—to tell GCP to route these types of requests to your private IP address).

Basically, unless otherwise specified, everything in GCP is restricted to the local network.

11.2 Setting Up Your Database Server for Remote Access

In order to use our `template-node` as a database server, we will need to prepare the box for remote access to our database. To do that, we need to do the following:

1. Modify `/var/lib/pgsql/data/postgresql.conf` and set `listen_addresses='*'`. We don't *have* to set it to the private IP address specifically, because that is the only IP address we have anyway, and it won't be accessible from the Internet unless we configure GCP to allow it. Restart PostgreSQL with `systemctl restart postgresql` so that the changes take effect.
2. Change the firewall so that we allow connections to the PostgreSQL server. This will be the commands `firewall-cmd --add-port 5432/tcp` and the same command with the `--permanent` flag attached.
3. Modify the PHP code `getReadOnlyConnection()` and `getReadWriteConnection()` so that they connect to the correct private IP address.

Now the server is ready to be used by a load-balanced cluster.

11.3 Creating a Replication Image

Creating a replication image in GCP is a somewhat overly complicated three-step process. First you need to create a "snapshot" of your instance, then you need to create an "image" out of that snapshot, and finally you will need to create an "instance template." The snapshot is essentially a backup of a machine. The image is essentially a snapshot that is intended to be used for creating new boot images for machines. Finally, the instance template combines an image together with machine settings (size, configuration, etc.) that can be used to very quickly deploy identical machines.

Creating a snapshot is a fairly straightforward process. From the main menu (click the three bars next to where it says “Google Cloud Platform”), go to the “Compute” section, choose “Compute Engine,” and then choose “Snapshots.” Click the “Create Snapshot” button. It will ask for a name for your snapshot, a source disk, and location. Name it whatever you wish, choose your existing VM Instance as the source disk, and you can leave the location alone (“Multi-regional” is the most flexible choice). Click “Create” and GCP will create a new snapshot for you.

Now, GCP does allow you to create instances out of snapshots. However, to make use of even more features of GCP, it is best to create what GCP calls an “image” out of your snapshot. In order to do this, go to the main menu, then the “Compute” section, choose “Compute Engine,” and then choose “Images.” It will load with a huge list of preconfigured images. You can ignore those. We want to make our own image. Click the “Create Image” button to get started.

Give the image a name (e.g., `template-image`) and set the “Source” to be “Snapshot.” This will bring up a menu asking which snapshot you want to create the image from. Choose the snapshot that you just created. If you want, you can add in a “Family” name. This will enable you to create updated versions of this instance which carry the same “Family” name.

Now click “Create.” From here you can create new machines from the image. When creating a new machine, under “Boot Disk,” your image will be available under the “Custom Images” tab.

Finally, we need to package this image into an instance template. Instance templates are found under the main menu under “Compute Engine” and “Instance Templates.” The process for creating an instance template is the same as for creating a regular VM Instance. The difference is that it will not immediately create any instances, but instead can be used later to quickly deploy fully preconfigured machines. Be sure when creating your instance template that you set the Boot Image to be the image you just created in the previous step

☞ BEING MORE CAREFUL WITH OUR TEMPLATE

Just as a note, because we set up the database on the machine that became our template, each new web server in the cluster will actually have an unused copy of the database with PostgreSQL running on it. This isn't a problem per se, but if you were going to make a production deployment, you would probably want to be sure that PostgreSQL was turned off on the template. Since GCP has a lot of steps anyway, the goal of this chapter was to reduce the number of steps you had to achieve in order to get a running configuration.

11.4 Creating Load-Balanced Groups

However, you can also create an automatically scaling, load-balanced group of machines called an "instance group." This is similar to what we did in Chapter 5, but GCP will actually manage scaling your app for you. In other words, as the load on your machines increases, GCP will automatically boot new, identical machines and add them under the load balancer.

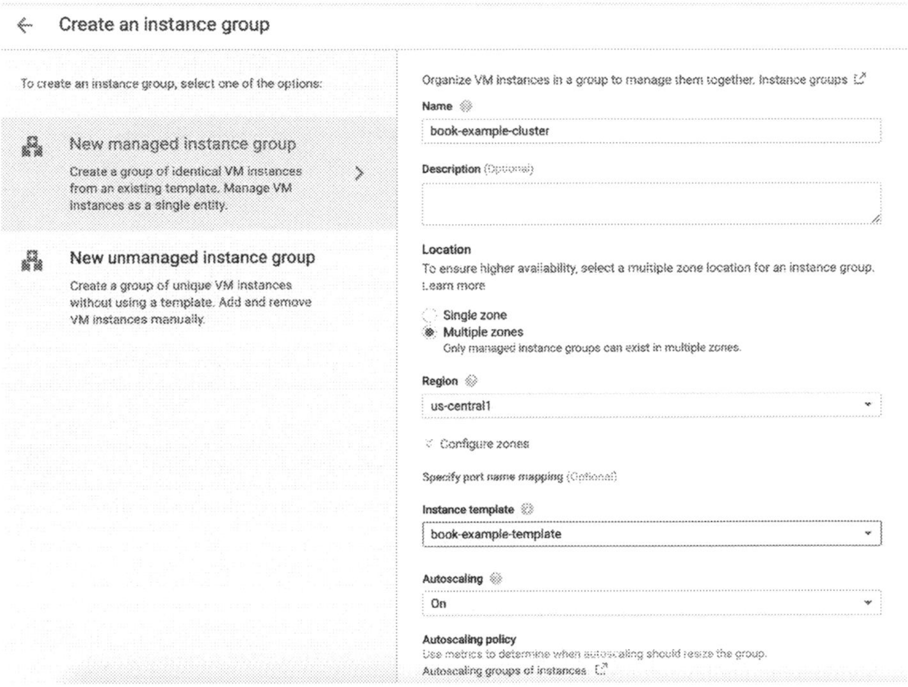

Figure 11-3. *Creating an Instance Group*

You can find instance groups on the main menu under "Compute Engine." Click "Create Instance Group" to start the process. Figure 11-3 shows what the process looks like.

Give your instance group a name. To ensure increased failure tolerance, select "Multiple Zones" under Location. Select the instance template that we created in Section 11.3. Be sure that "Autoscaling" is set to "On." If you like, you can set the minimum number of instances to be higher than one, in order to make sure that GCP is load balancing your app across multiple servers. Click "Create" to build your instance group.

By default, the instance group does very little. What we need is a way to bring traffic to the instance group. This is done with a load balancer.

To create a load balancer, under the main menu, look for "Network Services," and then "Load Balancing." Click "Create Load Balancer" to get started. Next choose "HTTP(S) Load Balancing." Next choose "From Internet to My VMs" since we want this load balancer to serve as the gateway between the Internet and your machines.

The next screen gives you the main areas of configuration. First, give your balancer a name. Next, under "Backend Configuration," select "Backend Services" and then "Create a Backend Service." You will need to name your backend service (the name doesn't really matter), then choose your instance group, and then create a health check (the health check just needs to be set to TCP port 80). Then click "Create" and it will create your backend service. You can leave "Host and path rules" and "Frontend configuration" to their default values. Click "Review and Finalize" to see all of your settings. Finally, click "Create" to build your load balancer. You can access the generated IP address, and it will balance the load across your machines.

Keep in mind that GCP takes quite a while to actually finish creating your load balancer. Even after GCP "thinks" that it is all the way created, and tells you that all of the instances have been added to the load balancer, it still takes several minutes for this actually to be the case. So, for the first several minutes that your load balancer is active, GCP may report errors when accessing URLs.

REMOVING LOAD BALANCERS

Removing a GCP load balancer` is harder than it seems it should be. To remove a load balancer that we have created in this way, we need to do the following:

1. Delete the load balancer itself from the list of load balancers.
2. On the load balancer listing screen, there is a tab called "Backends." Click that to view your backend services.
3. Under the "Backends" tab, click the backend you created for your load balancer, and delete that.

4. Now your balancer is removed, but you also need to get rid of your instance groups as well, or else you will continue to be charged for your machines (you won't be able to remove your instances or instance groups before the other steps are done).

5. If you want to not be charged for your image, you will also have to delete your instance template, and then delete your image, as well as your snapshot.

11.5 Other GCP Services

GCP, like AWS, has a number of other services that can be useful when creating cloud applications. Several similar services include

- Cloud SQL (similar to RDS)
- Storage (similar to S3)
- Memorystore (similar to ElastiCache)

Additionally, GCP also provides PaaS services through a service called Google App Engine.

In all, GCP has many of the same services as Linode and AWS, but they are a little more complicated to use. In certain extreme cases, that complexity can add to additional configurability, but it is rarely needed. As an example, GCP makes it relatively straightforward to map different subdirectories onto different instance groups. This makes it easier to host multiple applications under the same hostname and have each application be a different instance group.

For most applications, GCP's complexity outweighs the configurability that it offers.

CHAPTER 12

Server Management Techniques

So far, any time we have wanted to push an update to our cloud, we have had to reimage all of our servers. That isn't terrible, but neither is it the best option. If you imagine running a server farm with 20 different servers, do you really want to go and reimage them all every time you redeploy code? Probably not. If you want to install a server patch, do you want to go and reimage all your servers? Again, probably not.

Thankfully, there are a number of tools which can aid the management of a number of machines. They all have a variety of different focuses—some are focused on just making repetitive tasks easier, while others are full management solutions. I'm a simple guy, and I usually prefer the simpler tools to the large, all-encompassing ones.

12.1 Running Commands on Multiple Servers

Let's say that we wanted to install a new piece of software on every machine in the cluster. ImageMagick is a popular package for image manipulation. If we wanted to install ImageMagick on all of the nodes, the process is rather straightforward. We just ssh in to each box, run `yum install -y ImageMagick`, and then exit.

J. Bartlett, *Building Scalable PHP Web Applications Using the Cloud*,
https://doi.org/10.1007/978-1-4842-5212-3_12

Thankfully, there is a piece of software that will do that for us, called `pssh`, which stands for Parallel SSH. You don't need to install anything new on the server. `pssh` installs on your local machine and then just uses the regular `ssh` mechanisms for running the same command on all of the remote machines.

However, it might be simpler to install `pssh` on your `template_node` machine so you can use it from anywhere (i.e., you can log in to your template node and run `pssh` from there). `pssh` is available in the EPEL packages, so we can install it onto `template_node` just by doing the following:

```
yum install -y pssh
```

To use it, we need to create a file (we will call it `servers.txt`) that lists all of our servers that we are managing. Use `nano` to create the file, and put the *public* IP addresses of each server in the file. To use `pssh` all you have to do is type:

```
pssh -A -h servers.txt --user root COMMAND
```

Just replace `COMMAND` with the command you want to run. Therefore, we can issue the following command to install ImageMagick to each server:

```
pssh -A -h servers.txt --user root yum install -y ImageMagick
```

`pssh` will ask us for the password (that's what `-A` does), and then it will let us know when it has completed its tasks. Now all you have to do is make sure you keep the `servers.txt` file up to date, and you can easily perform most management tasks from a single machine.

When the commands are run, it will tell you the status on each server, and, if there is a failure, it will tell you the error code given by the command. For example, for three servers, the output could look like the following:

```
[1] 23:00:32 [SUCCESS] 45.79.7.179
[2] 23:00:32 [SUCCESS] 45.79.7.180
[3] 23:00:32 [FAILURE] 45.79.7.181 Exited with error code 1
```

Then you would want to look into the machine(s) that failed the command and determine what went wrong.

12.2 Syncing Files on Multiple Servers

Since running a command on every server is now easy, it would be nice to be able to copy a set of files to every server. Thankfully, pssh comes with two file-copying utilities called pscp and prsync so that you can easily copy files out to remote servers. That way, when you deploy a new version of your software, you won't have to reimage every server. The pscp program is sometimes called pscp.pssh, so if after installing pssh the command pscp is not found, try using pscp.pssh.

Let's say you had the file testme.html and you wanted it to be copied to /var/www/html on every server as user fred. To perform that task, just type the following (all on one line):

```
pscp -A -h servers.txt --user fred
 testme.html /var/www/html/testme.html
```

If you want to copy a whole directory, you can use the same process, but add -r for recursive copying.

One issue with pscp, however, is that it won't delete files for you. If you want to keep two directories in full sync with each other, for both additions and deletions, you can use prsync, though its syntax is a little tougher.

If I want to mirror a local directory mirror_me to the servers at /var/www/html/mirror_me, I would issue the following command (all on one line):

```
prsync -A -a -x --delete --user fred
 -h hosts.txt mirror_me /var/www/html/
```

By using pssh for commands, pscp for individual files, and prsync for entire directories, it is fairly easy to do basic management of a set of servers.

There are also several tools available specifically for application deployment on multiple servers. One of the more popular of these is Capistrano. Capistrano is written and customized in Ruby, though it can deploy applications in any language. Capistrano automates many tasks concerned with file deployment including

- Synchronizing with `git` repositories
- Maintaining previous versions of the application on the server for fast rollbacks
- Using symlinks to manage the deployment so that the entire deployment switches over in an instant
- Running custom tasks and scripts associated with the deployment

As your app gets more and more complex, your tooling needs will become more complex as well, and having a tool such as Capistrano will satisfy many of these needs.

12.3 Full-Service Solutions

While `pssh` allows you to issue commands to a group of machines and Capistrano allows you to automate deployments in a more robust manner, there are additional solutions that go further and completely manage the target systems for you. These are known as "configuration management" systems. These systems, while they handle a wide variety of situations, also come with complications. For most systems, I think that configuration management systems are overkill, and they add more complexity than they solve. However, one thing that they do well is to force you to document what your configuration is (and hopefully why it is set that way). Simply having a properly configured server does not communicate to future admins (or yourself in the future) which parts of your configuration were set by default and which parts are specifically configured for a purpose. With configuration management tools, your system configuration can be both documented and version controlled.

As someone who likes minimalist approaches, for system configuration I really appreciate the Ansible system, which does not even require anything to be installed on the remote servers. Ansible does all of its configuration via `ssh`, and is fairly easy to get up and running. There is a nice UI available from `http://ansible.com/` but it comes with additional monetary costs (though there are open source GUIs available as well). However, if all you need is a command-line tool, the open source version has one of the most minimal footprints of all of the configuration management tools.

You can even install it from the EPEL repository just with a simple:

```
yum install -y ansible
```

However, if you want to go full scale, while there are several other choices available, the one that many developers choose is Chef. Chef is based on the Ruby programming language, and lets developers make configurations as complicated as they wish. Chef also has a number of feedback mechanisms which allow you to view the status of all of your managed servers. With Chef, not only can you configure servers however you wish, you can also collect data and analytics on them as well.

For applications just getting off the ground, I suggest just keeping a manual log of configuration changes. As your application grows and matures (and hopefully gets millions and millions of users), being explicit about your configuration management becomes more important, and it might be worthwhile to move to a more all-encompassing solution.

Some of the cloud systems we have looked at have some amount of configuration management built in. GCP's Instance Template/Instance Group system can be thought of as a minimalistic configuration management system. Essentially, as you keep your instance template up to date, GCP will deploy it across your instance group. Elastic Beanstalk allows customization of the machines that it deploys your application on using environment customizations called “EB Extensions.” Even though Elastic Beanstalk is technically a PaaS, these extensions allow you to perform configuration management on the platform that you are deploying to.

CHAPTER 13

Linux Security Basics

This chapter is not meant to be a complete guide to cloud security, but merely to point you in the right direction. Security is more of a process and way of thinking than it is a series of steps, so this chapter will focus on how you need to think to protect your servers' integrity and your data's integrity.

13.1 The Basic Considerations

The most important security considerations for system administrators are the following:

- How can I mitigate the risks of being connected to the Internet?
- What is the *minimum set* of services I can run on my system and still have it function?
- How can I limit access to the server for services that are not needed by everyone?
- Is all of my software patched with the latest security patches?
- What are ways that someone could abuse the services that are deployed on the server, and how can I mitigate that?

J. Bartlett, *Building Scalable PHP Web Applications Using the Cloud*,
https://doi.org/10.1007/978-1-4842-5212-3_13

The key to security is to identify what the server is used for and then to remove anything that does not directly contribute to that function in order to prevent unwanted consequences. Every service that you have running is a potential security hole—it is something that someone may one day find out how to exploit. Therefore, you don't want exposure on points that do not contribute directly to function.

13.2 Examining Your Current Server

The best tool for checking what services are currently running and listening on your server is `ss` (socket statistics). To find every TCP service that is listening for a connection, run the following command as root:

```
ss -plnt
```

Every line that says `LISTEN` is a service that is listening for an IP connection. If the "Local Address" is `127.0.0.1`, `::1` (for IPv6 addresses), or the node's private IP address, that means that the service is protected—only applications *on the machine itself* can see the service (on on the private network, in the case of the private IP address). However, if the local address is `*`, `0.0.0.0`, or `::` (for IPv6 addresses), that means that it is available for anyone to connect to.

Similarly, for UDP connections, you can do:

```
ss -plnu
```

For each service (TCP or UDP) that is available to connect to, you should be sure that it is not permitted by your firewall unless you are absolutely sure you want people connecting. To get a list of services that your firewall allows, issue the command:

```
firewall-cmd -list-all
```

The items that are listed under "services" and "ports" are what the firewall is letting through. The services that you should allow are `dhcpv6-client` (used by the cloud for configuring networking), `ssh` (so you can log in), `http` (for non-SSL HTTP connections), and `https` (for SSL-enabled HTTP connections).

Even though the firewall will prevent connections to anything not listed, if you have any service running that is listening for connections that you don't specifically need running, you should disable the service. Additionally, you should periodically check your firewall to make sure it is configured correctly.

Additionally, you should take inventory of all of the running processes, even if they aren't listening for connections. The best way to do this on Linux is using the command `ps -afxww`. Note that everyone has their own favorite way of calling `ps`, but this is mine.

I can't give you a list of everything that should/shouldn't be running. However, it is best to learn what each piece does and turn off the pieces that you don't need. The fewer services running, the better.

You can also list out all of the packages that you have installed using the command `rpm -qa`. As you become more experienced, you should be able to uninstall programs that you don't need.

13.3 The Root User

The most dangerous part of a Linux installation is the root user. It is dangerous for two reasons. First, it is the superuser, so it has power over everything else. Second, everyone knows its username, which makes it easier for automatic hacking tools to breach.

There are several ways to mitigate problems with the root user. They include

1. The root password should be secure (i.e., difficult even for a computer to guess). In fact, the passwords of *every* user should be secure.

2. The server should not permit remote direct root logins. To prevent direct root logins via `ssh`, modify `/etc/ssh/sshd_config`. If there is already a line that says `PermitRootLogin`, change the value from `yes` to `no`. If there is not a line there, add a line that says `PermitRootLogin no`. After saving the file, do `systemctl restart sshd` and the change will take effect. After this, you will need to log in as an ordinary user and use the `su` command to switch user to `root`.
3. Services should rarely run as the root user unless there is an overwhelming reason to do so. If a service must do something as root, the part of the service which directly talks to remote computers should not be root.
4. Users should not spend much time as the root user. In this book, most of our time has been spent as the root user. Instead, users should log in under their own accounts and then switch to the root user using `su` or `sudo` for temporary root privileges.
5. You should install some sort of service denial program such as `fail2ban`, which disables logins from particular IP addresses after a certain number of failed attempts.

By preventing people from becoming the root user, running services that don't run as the root user, and by protecting the root account from outside access, the ability of an intruder to do damage is greatly diminished.

Most administrators use `sudo` to manage access to the root user. The `sudo` command gives a user temporary access to the root user *using their own password*, or without a password once logged in. It is already installed on your machine and will allow any user that is in the wheel group to run any command as root. To add a user (say, fred) to the wheel group, issue the command:

```
usermod -a -G wheel fred
```

Now, the fred user can run any command as the root user by just prepending sudo to the beginning of the command. For instance, if fred wanted to display the file /etc/sudoers (the configure file for the sudo command), Linux would not normally let him. If he did cat /etc/sudoers, the operating system would give him an error. But, if fred is in the wheel group, and he issued the command sudo cat /etc/sudoers, the operating system would ask him for his password and, after reauthenticating him, would run the command for him.

In order to use pssh with PermitRootLogin=no, you will need to use sudo to switch users. However, pssh doesn't like interaction and will cause problems when the user is asked for the password. Therefore, you need to modify the default configuration of sudo in order to allow the user to utilize sudo without supplying a password (they will still need their password to log in). To do this, as the root user, add the following line to /etc/sudoers:

```
%wheel ALL=(ALL) NOPASSWD: ALL
```

Now, to use pssh, you would log in as fred and sudo to perform your system administration command, like this:

```
pssh -A -h servers.txt --user fred sudo put_your_command_here
```

13.4 Installing a Web Application Firewall

A web application firewall is a piece of software that sits between your web application and the Internet. The purpose of a web application firewall is to check incoming traffic for patterns that are known to be consistent with malicious intent, and then block those requests. A web application firewall does not make up for bad programming in a web application, but it will often prevent automated hacking tools from finding holes.

Apache has a web application firewall available for it that is easy to install. To install it, just do the following as root:

```
yum install -y mod_security mod_security_crs
systemctl restart httpd
```

However, with our test site, the web application firewall will likely block requests that use the IP address in the URL, as that is a characteristic of many hack attempts! In fact, I have found that for many production systems I have to disable several individual rules to get the web application firewall to work with my application. This can be painful, but on the whole it is worthwhile to do.

The web application firewall will log which rule denied the request in the log file `/var/log/httpd/error_log`. Therefore, you can find the rule ID number in the log and then disable it in the configuration. Just add the line `SecRuleRemoveById IDNUM` to the file `/etc/httpd/conf.d/mod_security.conf` to disable unwanted rules.

13.5 Checking for Rootkits

A *rootkit* is a piece of software installed by someone who has broken into your server that makes it easier to control your server for nefarious deeds. If your server is secure, it is unlikely that someone will break in, but nonetheless it is good to periodically check.

The two standard pieces of software for rootkit checking are `rkhunter` and `chkrootkit`. `rkhunter` is currently a part of EPEL and can be installed with:

```
yum install -y rkhunter
```

To run the program, just do `rkhunter --check`. Be aware that it can generate false warnings and false positives, so be sure to check the logs to see what, specifically, it found when it was looking at an issue.

13.6 Other Security Software

There are a host of other security packages that you can install and use. The important thing is to know what measures are available, and to see whether or not they are cost-effective for your needs.

Some additional common security packages for Linux include

logwatch: This tool analyzes log files and e-mails administrators when suspicious activity is recorded.

fail2ban: This program looks for repeated login failures and other suspicious behaviors and will block IP addresses that look like they are attempting to break in.

SELinux: Security-Enhanced Linux is an operating mode where the Linux kernel gives very fine-grained access control to programs, limiting what each program and user can do significantly more than normal. SELinux can provide a lot of risk mitigation, but it is fairly complex to set up, and it is easy to accidentally block your own applications from doing what they need to do. We have disabled SELinux in this book because of the amount of configuration issues it entails.

FirewallD: This is the standard firewall administration application used in this book.

AuditD: This program looks for and logs suspicious activity by application programs.

Remote Syslog: This is not a program, but the system logger can be configured to log to a remote server, so that intruders cannot cover their tracks by modifying log files.

13.7 Application Security

The hardest thing to secure is the application itself. Realistically, I can't offer a whole lot of tips without writing another book. Nonetheless, the most important things to remember are

1. Code defensively.
2. Verify every piece of data from the user.
3. Properly escape everything sent to the user.
4. Always double-check the privilege of a user before performing an operation or showing data. Make sure that simply knowing a URL or a parameter doesn't automatically give a user undue power.
5. Be very careful about anything that is sent to an external command or program. Double-check that you have properly escaped or filtered everything.
6. Always imagine what would happen for every piece of data and every request if someone were maliciously manipulating it.

7. Always try to code using "best practices" (e.g., `https://phptherightway.com/`). Using best practices when coding can save you from security problems that other people introduce, including those that you accidentally introduce yourself later down the line as your application grows and turns what is currently safe code into unsafe code due to changes elsewhere in the codebase.

Also be sure to check the list of common vulnerabilities at `www.owasp.org`

There are many other things that you can and should do to secure your server and your application, but hopefully this has given you a starting list of things to be thinking about. PHP sometimes gets an unwarranted bad reputation for security problems. However, the reason for it is not the language, but rather that it is often the first language learned by newer web programmers who have less security experience.

A few resources to help you get started in this direction include

1. ***Securing PHP Apps* by Ben Edmunds**: This is a short and to-the-point guide to secure practices in PHP.

2. ***Pro PHP Security* by Chris Snyder, Thomas Myer, and Michael Southwell**: This is a more comprehensive guide to security and security principles with a PHP focus. It is an older book, but the core security principles have not changed.

3. ***Mastering Linux Security and Hardening* by Donald Tevault**: Writing secure PHP code won't help you if your server isn't configured properly.

4. ***Practical Information Security Management* by Tony Campbell**: This book will help you understand at a higher level what is being secured, what is being protected, how to manage tradeoffs, and what the ultimate goals of security are.

The most important thing, however, is to always be thinking about how your application could be abused and always be learning new ways to proactively guard against those things.

APPENDIX A

List of Linux Commands

This appendix contains lists of the Linux commands that were used in this book, plus a few extra that are frequently important when working with Linux and the CentOS Linux distribution.

Linux commands usually follow the same basic format, based largely after human communication. The command is listed first because that is the name of the program which runs the command. Most commands take one or more *arguments* which can be considered similar to direct objects of a sentence. Additionally, many commands take options (also called flags, modifiers, or parameters) which modify the behavior of the command. These are very similar to adverbs. Options are usually given by prefixing them with a dash. For instance, to change the "list directory" command (`ls`) to show a long-form listing, you add the `-l` option to the command. Thus, the command to show the long-form listing is `ls -l`.

Sometimes options themselves take arguments. However, if they don't, you can oftentimes stack them. For instance, `ls` can also show hidden files with `-a`. You *can* do both options with `ls -l -a`, but you can also squeeze them together with `ls -la`. Many commands have both long options and short options, with the longer version of the option usually preceded by

J. Bartlett, *Building Scalable PHP Web Applications Using the Cloud*,
https://doi.org/10.1007/978-1-4842-5212-3

two dashes. For instance, instead of doing `ls -a`, you could instead type `ls --all`. If you don't know how to invoke a command, most commands have a help screen that you can invoke by doing `COMMAND -h` or `COMMAND --help`.

While each command has its own quirks and abilities, this general command format is common to most commands in Linux.

Additionally, you can get fuller information about a command by looking it up in the manual using the `man` command. For instance, to get the manual page on `ls`, type in `man ls`. Type `q` to exit the manual pages.

Basic Linux Commands

`cat`: This command simply spits out a given file to the screen.

`cd`: This changes your current directory. If the argument starts with a `/`, then it is an *absolute* path starting from the root of the filesystem hierarchy. If the argument starts with a `~`, it is a path that starts from your home directory. Otherwise, it is a *relative* path starting from your current directory. The special directory `..` represents the parent directory, and the special directory `.` represents the current directory. Therefore, to get to the parent directory of your current directory, just do `cd ..`.

`chgrp`: This is like `chown`, but just for group ownership.

`chmod`: This changes the permissions on a file. Each file has separate permissions for the user owner (`u`), the group owner (`g`), and everybody else (`o`). The basic set of permissions is read (`r`), write (`w`), and execute (`x`). To add execute permission to the owner of a file, you would type `chmod u+x FILENAME`. To add read permission to everybody on a file, you would type `chmod a+r FILENAME`, where a refers to all users (u, g, and o). To remove group write permission on a directory and every file/directory under it, you would do `chmod -R g-w FILENAME`.

`chown`: This changes the user and group ownership of a file. If you do `chown fred FILENAME`, then `fred` becomes the owner of the file. If you do `chown fred:stonemasons FILENAME`, then `fred` becomes the owner and the group `stonemasons` becomes the group owner of the file. You can add the `-R` option to change the ownership of a directory and all of the file/directories underneath it.

`head`: This command gives you the first few lines of a file. `head -n NUM FILENAME` will give you the first `NUM` lines of the `FILENAME` file.

`ls`: This command lists all of the files in your current directory. Using the `-a` switch will show all files, including special and hidden files (files starting with a `.`), and the `-l` switch will show additional information about the files. You can also specify a specific directory to list at the end of the command.

`man`: This command gives the *man*ual page for a given command or configuration file. `man ls` tells you about the `ls` command. To get out of the manual, just type `q`.

`mkdir`: This creates a new directory within your current directory (or anywhere at all if you give it an absolute path starting with / or ~).

`nano`: This is a simple text editor that comes with many Linux distributions. Common commands to use within `nano` are control-o which saves (outputs) the file, control-x which quits, and control-w which searches.

`pwd`: This prints out your current directory. It stands for "print working directory."

`rm`: Removes a given file. If you want to remove a whole directory tree (the directory and all of the files/directories in it), add the `-r` switch. Just be careful!

`scp`: This command (known as "secure copy") remote copies a file from one machine to another over an encrypted channel. The basic format is `scp LOCALFILE USERNAME@REMOTE.MACHINE.NAME:/PATH/TO/DESTINATION`.

`ssh`: This command (known as "secure shell") allows you to remote access other machines over an encrypted channel. The general format is `ssh` USERNAME@MACHINE.HOST.NAME.

`tail`: This command gives you the last few lines of a file. `tail -n NUM FILENAME` gives you the last `NUM` lines of the file `FILENAME`.

`telnet`: Telnet used to be the way that you accessed machines remotely, before encryption became a necessity. Now it is often used to make direct connections with remote servers for testing. For instance, to talk to a web server directly, you can type in `telnet REMOTE.MACHINE.NAME 80`, and it will connect you directly to port 80 on the remote machine. Remember that `telnet` will display information directly to your screen, so be careful if sensitive data may be returned.

`vim`: This application (or its older brother, `vi`) is an editor that you will find on nearly *any* Linux-like system. It is very powerful, but please read a tutorial on it before attempting to use it. Its primary benefits are that its keyboard interface is based on where your hands naturally sit on the keyboard and that its small footprint and longstanding heritage mean that you will never be on a Linux or UNIX machine that doesn't have it installed.

Basic System Administration

`passwd`: This command changes the password of users. Without an argument, this changes your own password. Otherwise, it changes the password of the user you specify.

`useradd`: This creates a new user on the system. `useradd fred` adds a new user with the name `fred`. Use the `passwd` command to set their password.

`usermod`: This modifies a user on the system, normally to add them to a group. Use `usermod -a -G thegroup theusername` to add the user `theusername` to the group `thegroup`.

`systemctl`: This command handles starting and stopping background services on Linux. This command usually has the form `systemctl CMD`

SERVICE where CMD is the command you want to give, and SERVICE is the service you want to issue the command to. Service commands include start, stop, restart, enable (make sure the service starts on bootup), and disable (make sure the service does not start on bootup). The main services covered in this book include httpd, postgresql, and memcached.

firewall-cmd: This command handles the firewall. This command has several options. The option --add-service SERVICENAME allows you to open up a service to outside connections, where SERVICENAME is the specific service you want to allow outside users to connect to. The list of services is available by running firewall-cmd --get-service. If you just want to open up a port (i.e., one for which there is no service description), you can just use firewall-cmd --add-port 1234/tcp in order to open up TCP port 1234. To make the service available on reboot, you need to re-issue the command with the --permanent flag added. You can show everything enabled on the current firewall by using --list-all.

yum: The yum command is the automated package installer for CentOS and other Linux distributions. yum allows you to quickly and easily search, find, and install Linux packages onto your system. yum focuses on finding packages on the Internet and resolving dependencies between packages, and then uses rpm to do all of the dirty work of actually installing the packages. yum includes several subcommands, such as `search`, `install`, `update`, and `uninstall`. `yum update` updates all of your installed packages, `yum search TERM` gives you a list of all available packages whose description includes the word `TERM`, and `yum install PACKAGENAME` will install `PACKAGENAME` and all of its dependencies for you. If for some reason `yum` stops working correctly, usually you can fix it by running `yum clean all`.

`rpm`: The `rpm` command is the low-level package manager for CentOS. It handles the work of actually taking a package file and installing it onto the system. This is pretty rare, as this is usually handled through `yum`. However, `rpm` also has a way of querying installed packages. A list of all of your installed packages can be found by running `rpm -qa`, and a list of all files that were modified after installation can be found by running `rpm -Va`.

`rkhunter`: This command, if installed, checks your system for various types of malware by running `rkhunter --check`.

`su`: This command stands for "switch user." Without any arguments, this switches the user to the `root` user. You usually want to add the `-l` option, which means to act as if you logged in with this user, which will take you to the user's home directory and run other login tasks. If you give it an argument, it will be the name of the user you want to switch to. You must enter that user's password in order to switch users.

`sudo`: This command lets you temporarily run a command as another user (normally as `root`). The configuration of this command is beyond the scope of this book, but `man sudo` should give you good information.

`pssh`: This command performs a parallel `ssh` session across multiple hosts. See Chapter 12 for more information about this command.

`pscp`: This command performs a parallel `scp` copy from a local file or directory to multiple destination hosts. See Chapter 12 for more information about this command.

`prsync`: This command does a parallel synchronization between a local directory and multiple destination hosts. See Chapter 12 for more information about this command.

`ss`: This command gives information about open sockets on your machine. The two commands we focus on are `ss -plnt` for looking at listening TCP connections and `ss -plnu` for looking at listening UDP connections. This command is critical for knowing potential attack vectors that an attacker may use to gain access to your system.

`netstat`: This is an older version of the `ss` command. This command gives you lots of information about active network connections on the system. The two ways this is normally called are `netstat -plant` (which gives a list of TCP session and listening sockets) and `netstat -planu` (which gives a similar list for UDP).

`ps`: This command gives you information about processes running on the system. This has numerous options that can give you almost any piece of information you want to know. However, my favorite way of calling it is `ps -atxww` which gives you a list of all of the processes currently running displayed as a tree so you know which process spawned which other process.

`top`: This command gives you information about which processes are using the most system resources. Use `q` to leave `top`.

`free`: This gives a short rundown of the current memory usage on the system. `free -h` gives the most readable output.

`uptime`: This gives a short rundown of the current load on the system. In Linux, the load is the number of processes that are wanting CPU time at any given moment. Therefore, for a machine with *x* processors, the machine is fully loaded near *x* and is falling behind when it goes above that number. I usually try to keep my machines only half loaded at most.

PostgreSQL Commands

`createdb`: This command creates a new database in PostgreSQL, usually called with the parameter `-U PGUSER`, where `PGUSER` is the PostgreSQL user who will create (and therefore own) this database.

createuser: This command creates new users in PostgreSQL. This takes the -U PGUSER parameter to tell which user to run as. The -d flag will give the new user permission to create new databases, and the -P parameter will prompt you to type in a password for the new user.

pg_basebackup: This command creates a binary backup of PostgreSQL. In this book, we use this command as a starting point for replication.

postgresql-setup: This is a script that aids the installation of PostgreSQL instances. In this book, we just use postgresql-setup initdb to create the initial instance data for PostgreSQL.

psql: This command gives you access to the PostgreSQL interactive SQL prompt. The -U PGUSER indicates which PostgreSQL user you will access the database as, and the command argument will be which database to connect to. The -h HOSTNAME option can allow you to access a PostgreSQL database on a different host.

Other Application-Specific Commands

ab: The ApacheBench command simulates a large number of requests to a web site and gathers response statistics. It is normally called ab -n NUMREQUESTS -c NUMCONNS FULL_URL, where NUMREQUESTS are the total number of requests for ApacheBench to make, NUMCONNS is the number of simultaneous connections to keep going, and FULL_URL is the destination URL you are trying to test.

`convert`: This command is a part of the ImageMagick package, and is a Swiss army knife for converting and modifying image files. To convert a JPEG file called `testme.jpg` to a PNG file, just do `convert testme.jpg testme.png`.

`pecl`: This is a package manager for PHP. This allows you to install add-ons to the system's PHP environment that aren't available via yum.

APPENDIX B

Important Files and Directories

This book mentions a number of important files and directories on CentOS. This appendix lists the files and directories from the book that you need to remember, plus a few more. Remember, each distribution has its own special way of doing things, so a file's location on CentOS may be slightly different than its location on Ubuntu. Additionally, each cloud provider may also set things up in certain ways. Custom-installed applications sometimes can wind up just about anywhere, depending on the application author, the person who packaged it, or the user who installed it. The Remi packages used in this book are a case in point—the `/opt/remi` hierarchy is entirely an invention of the packager.

Basic Linux Filesystem Directories

The first part of knowing where to find things is to know where Linux likes to put things. This section is a brief introduction to the standard structure of the Linux filesystem. Note that on Linux hard drives do not exist separately, but are instead "mounted" at certain locations on the

J. Bartlett, *Building Scalable PHP Web Applications Using the Cloud*,
https://doi.org/10.1007/978-1-4842-5212-3

filesystem. In other words, there is a single filesystem hierarchy even with multiple hard drives. The drives simply represent specific folders within the hierarchy.

`/`: This is the root of the filesystem tree. Every file and directory is contained somewhere in here.

`/boot`: This directory holds basic files for booting up (like the Linux kernel). You should usually stay out of this directory.

`/dev`: In Linux, devices are represented as files, and they live here.

`/etc`: This directory contains most of the configuration files for the computer.

`/proc`: This directory contains a file for every running process on the computer, plus files to represent operating system status information.

`/tmp`: This directory is used to store temporary files.

`/var`: This directory holds *var*iable files—files that are tied to programs and are intended to change during the operation of a program. For instance, your database is stored in a subdirectory of `/var` because it is changing and it is managed by the database software instead of by the user.

`/var/log`: This directory holds most of the log files for the system.

`/var/spool`: This directory is mostly used for transient data within a system, such as current mail for a mail server, jobs for a print service, and so on.

/usr: This directory is a mostly read-only directory used to store programs used by the machine in the course of its operation.

/usr/bin: The files in this directory are the programs (i.e., *bin*aries) that are normally used on the server by users and other programs. Also note that there is a directory /bin which has the programs that are necessary for proper bootup.

/usr/sbin: The files in this directory are the system programs (e.g., daemons and system tools) that are available for use on this machine. Also note that there is a directory /sbin which has the system programs that are necessary for proper bootup.

/usr/lib: The files that live here are the system *lib*raries that support the applications. There is often a /usr/lib64 directory for 64-bit libraries. Also note that there is a directory /lib which has the libraries that are necessary for proper bootup.

/usr/local: This directory has almost an identical structure to /usr, but the programs installed here are usually compiled by the system administrator. Unlike /opt (where each program gets an entire directory to themselves), the programs installed here all share the same bin, lib, and sbin directories.

/home: This directory holds the home directory of each user on the machine except root.

/root: This is root's home directory.

`/opt`: This is a directory that often has custom-installed applications.

In `/opt`, each program usually has its own directory.

Important Directories for Cloud Servers

These directories are special-purpose directories that you will want to know about for operating a cloud server. Different Linux distributions may put these in different locations, but these are the default CentOS locations.

`/etc/httpd`: This is the directory that holds the configuration files for the Apache web server.

`/etc/postfix`: This is the mail server's configuration directory.

`/etc/sysconfig`: This directory holds additional configuration of many system services.

`/etc/systemd`: This directory contains the configuration information that managed the `systemctl` command.

`/etc/ssh`: This directory contains the configuration of both the `ssh` client and server.

`/var/lib/pgsql/data`: This is the directory that holds PostgreSQL's database files.

`/var/lib/pgsql/data/pg_log/`: This is the directory that holds PostgreSQL's log information.

`/var/log/httpd`: This is the directory that holds Apache's server logs.

/var/www/html: This is the default directory for holding a web site.

/opt/remi/php74/root: This is the directory that holds the PHP 7.4 installation from the Remi repository that is used in this book.

Important Files

This book has covered many different files that have to be configured for properly running a server.

/var/lib/pgsql/data/pg_hba.conf: This file configures access controls for PostgreSQL.

/var/lib/pgsql/data/postgresql.conf: This is the main configuration file for PostgreSQL.

/var/lib/pgsql/data/recovery.conf: This file controls a PostgreSQL instance running as a replica server.

/var/log/httpd/access_log: This is the default location that logs every time your web server is accessed.

/var/log/httpd/error_log: This is the default location for errors from Apache and PHP.

/var/log/maillog: This is the log of all mail messages sent out from the system.

/var/log/messages: This is the default location for system error messages.

APPENDIX C

What to Do When It Doesn't Work

There are all sorts of things that can go wrong when following code from a book. This appendix focuses on the most common issues and what you can do about them.

Making Sure Everything Is Typed in Correctly

In any programming book, the first thing to check when something doesn't go right is to make sure it is typed in correctly.

- Be sure that everything is typed correctly.
- Make certain that everything is on the line that it is supposed to be (sometimes this matters, sometimes it doesn't—better safe than sorry).
- Check to see that everything uses the right punctuation (i.e., that you used colons and semicolons properly and used single and double quotes correctly).

J. Bartlett, *Building Scalable PHP Web Applications Using the Cloud*,
https://doi.org/10.1007/978-1-4842-5212-3

- Verify that your computer didn't autocorrect your punctuation into something more pretty. If you type in `"hello"` and your computer spits out “`hello`”, then that will not work. Turn off automatic punctuation or whatever it is.
- If you are writing the PHP files locally, be sure that you are using a text editor, not a word process, and that the files are being saved text-only, using either an ASCII or UTF-8 character set.

Making Sure You Checked the Logs

If you think that you did everything right, and it still isn't working out for you, *check the logs*. This is the easiest way to spot a mistake. For PHP programs, the log files to check are `/var/log/httpd/error_log` and `/var/opt/remi/php74/log/php-fpm/www-error.log`. For PostgreSQL startup errors, the log files are in the directory `/var/lib/pgsql/data/pg_log`. Finally, for general system messages, check `/var/log/messages`.

Many log files are very long, and you only need to see the last few lines. The `tail` command will help you by just giving you the last few lines of a file. `tail /var/log/messages` or `tail /etc/httpd/logs/error_log` will potentially give you a lot of information about any problems you are having on your system.

Making Sure You Didn't Miss a Step

The instructions in this book contain a number of steps, and care was taken to be sure that they all worked when done in order. Therefore, be sure that you follow the steps in order when you are first learning. After you get the first thing up and running, *then* you should take the time and initiative to branch out and try variations.

Additionally, this book has a lot of things for you to install, and it is possible you missed one. Just in case, you might go each chapter and make sure you installed everything.

What If I Run a Different Version/Distribution of Linux

Of course, there are a ton of different Linux distributions, and the specifics even within a distribution change for each version. If you are running a different version of Linux, be aware that the commands might be slightly different, that there may be slightly different ways of doing things, different things might come standard, and directories might be located in different places.

This book was written around the latest CentOS distribution at the time, so it will hopefully be perfectly suitable for several years. Nonetheless, the basic ideas will still work even if you need to tweak the directions slightly for the Linux distribution you are working with.

What If I Want to Use a Different Cloud Service?

While this book focuses on Linode, AWS, and GCP, there is very little in this book that doesn't directly translate to other services. I am a huge fan of Linode because they have a service that is simple yet powerful, and their servers are top quality. Most cloud services, however, are structured in essentially the same way. Therefore, if you are using a different service for some reason, the basic ideas in this book should still continue to hold. The point of this book is to get your mind thinking about cloud *architecture*—using a specific service merely helps you get started with a concrete starting point.

Where Else Can I Find Information?

Information abounds on the Web about troubleshooting programs that have gone awry. The first try for anything that happens that I don't understand is to Google the error message. If that doesn't work (or if you don't even have an error message to start from), the next step is probably a message board. The biggest message board is Stack Exchange, but there are other good ones as well.

Each of the technologies that this book deals with has excellent reference manuals. Most of these are really good, though it sometimes takes a while to find the specific solution that you need. The PHP reference manual even has an interactive section, where developers can ask questions and get answers.

Finally, you can go to a local developer meetup. Nearly every city has one. Even if it isn't PHP specific, if you find a developer group, chances are one of them will know enough PHP, Apache, or Linux to help you out.

Don't be stuck by yourself—get help and improve your skillset!

Afterword

Now that you have taken our small guestbook app through a variety of cloud configurations, you should be able to apply these principles to scaling any other application in the cloud.

A few tips to keep in mind as you go forward are

- Always look for ways to restructure your application so as to prevent a bottleneck of any one location or action.
- Keep an eye out for parts of your application that can be easily replicated to infinite scale through CDNs or similar services.
- Measure the scalability of different configurations of your application to find out where your problems are and where you quickly max out your system.

If you have followed through the exercises in this book, you can consider yourself experienced at scaling web applications in the cloud!

J. Bartlett, *Building Scalable PHP Web Applications Using the Cloud*,
https://doi.org/10.1007/978-1-4842-5212-3

Index

A

B

C

J. Bartlett, *Building Scalable PHP Web Applications Using the Cloud*,
https://doi.org/10.1007/978-1-4842-5212-3

D

E

S

T, U

V

W, X

Y, Z

Printed in the United States
By Bookmasters